Erich von Däniken maintains that it is not enough merely to look at the oddities which archaeologists have not been able to explain away. We should go out looking for evidence of visits from extra-terrestrial beings. When, for example, will Tiahuanaco be properly excavated and the massive and technically advanced buildings attributed to a race other than the one whose bones and crude tools happen to have been found there? When will an international organisation catalogue and classify the thousands of rock drawings from all over the world, many of which seem to depict memories of space visits? What about the unexplored ruins deep in the jungles of Guatemala and Honduras? What about the Plain of Nazca? What about the Sahara?

'He plausibly and readably suggests that the earth has had extra-terrestrial visitors and that they have left definite evidence of their visits if only man will take the trouble to read it'—*Irish Press*

Also by ERICH VON DÄNIKEN

CHARIOTS OF THE GODS?

and published by CORGI BOOKS

# Return to the Stars

Evidence for the impossible

## Erich von Däniken

*Translated by* Michael Heron

**CORGI BOOKS**
A DIVISION OF TRANSWORLD PUBLISHERS LTD

# RETURN TO THE STARS

## A CORGI BOOK 0 552 09083 2

First published in Great Britain by
Souvenir Press Ltd.

### PRINTING HISTORY

Souvenir edition published 1970
Souvenir edition reprinted 1970
Souvenir edition reprinted 1970
Corgi edition published 1972
Corgi edition reprinted 1972 (Australian edition)
Corgi edition reprinted 1973
Corgi edition reprinted 1973
Corgi edition reprinted 1973
Corgi edition reprinted 1974
Corgi edition reprinted 1974

Copyright © 1968 Econ-Verlag Gmbh
Translation copyright © 1970 by Michael Heron
and Souvenir Press

This book is set in Granjon 11/12 pt.

Corgi Books are published by Transworld Publishers Ltd.,
Cavendish House, 57–59 Uxbridge Road, Ealing,
London, W.5.

Made and Printed in Great Britain by
Richard Clay (The Chaucer Press), Ltd., Bungay, Suffolk.

# CONTENTS

# ILLUSTRATIONS

Close-up of a hole.
Aztec ceremonial disc of serpentine.
Milestone of King Melichkhon.
Rock painting from Auanrhet, Tassili, about 8,000 years old.
The great Martian God.
Space travellers from a rock drawing in Val Camonica (Italy).
Rock painting found 25 miles south of Fergana (Uzbeck, USSR).
Circular Mayan calendar.

# ABOUT ERICH VON DÄNIKEN

Erich von Däniken is not a scholar. He is an autodidact, which the dictionary defines as a man who is self-taught. Probably this helps to explain the success his first book met with all over the world. Completely free from all prejudices, he had to demonstrate personally that his theses and theories were not unfounded and hundreds of thousands of readers were able to follow him along the adventurous road he took—a road that led into regions that were surrounded and protected by taboos.

Besides, his fearless questioning of all the previous explanations of the origin of the human race seems to have been long overdue. Erich von Däniken was not the first man who dared to challenge them, but his questions were more impartial, more direct and more audacious. In addition, he was able to say exactly what he wanted to say, unlike a professor, for example, who would have felt bound to take the opinions of his colleagues or the representatives of similar academic disciplines into consideration. What is more, he came up with some startling answers.

Men who bluntly ask bold questions that cast doubt on time-honoured, accepted explanations have always been a nuisance and people have never been over-fussy about how they silenced them. In the past their books were banished to secret libraries or put on the Index; today people try to hush them up or make them look ridiculous. Yet none of these methods has ever succeeded in disposing of questions which concern the reason for our very existence.

Erich von Däniken has the spontaneity of the enthusiast. In the summer of 1968 he read articles by Vlatcheslav Saizev in the Soviet journal *Sputnik* with titles such as 'Space-ship in the Himalayas' and 'Angels in Space-ships'. Von Däniken booked a flight to Moscow on the spot. There Professor Shklovsky, Director of the Radio-Astronomic Department of the Soviet Academy of Science's Sternberg Institute, answered his questions.

The author of *Chariots of the Gods?* was barely nineteen years old when his curiosity first drove him to Egypt where he hoped to track down the real meaning of some cuneiform inscriptions. Since his first journey in 1954, he hops on planes to clear up his theories the way we catch a bus. Thinking on the space scale as he does, distance means nothing to him so long as the goal of his journeys provides arguments for the impossible.

WILHELM ROGGERSDORF

# FOREWORD

Return to the stars?

Return? Does that mean that we came from the stars?

The desire for peace, the search for immortality, hankering for the stars—all these are deeply rooted in the human consciousness and have been ceaselessly pressing for realisation from time immemorial.

Is this urge for realisation that is so deeply implanted in human beings something to be taken for granted? Is it really only a question of human 'desires'? Or does this striving for fulfilment, this nostalgia for the stars, conceal something quite different?

I am convinced that our longing for the stars is kept alive by a legacy bequeathed by the 'gods'. Memories of our terrestrial ancestors and memories of our cosmic teachers are both at work in us. Man's acquisition of intelligence does not seem to me to have been the product of a long and tedious development. The process took place too suddenly for that. I think that our ancestors received their intelligence from the 'gods', who must have possessed knowledge that made the whole process a rapid one.

Obviously we shall not find proofs of my assertion on the earth if we stick to the existing methods of archaeological investigation. If we do, we shall simply and inexorably increase the existing collections of human and animal remains. Each find will be given its catalogue number, put in a glass case in a museum and kept clean by the museum staff. But we cannot approach the heart of the matter with such methods alone. For the heart of the matter, I am convinced, lies in the important questions of when and how our ancestors became intelligent.

This book is an attempt to provide new arguments for my theory. It is meant to be another peaceful incentive to reflection about the past and future of mankind. For too long we have failed to investigate our remote past with daring and imagination. It will not be possible to produce the last conclusive proofs in one generation, but the walls which still separate fantasy from reality will have more and more breaches in them. I shall try to do my best to keep on breaking through them with new aggressive questions. Perhaps I shall be lucky. Perhaps questions of the kind that are also asked by Louis Pauwels, Jacques Bergier and Robert Charroux will be answered in my lifetime.

I should like to thank the countless readers of my *Chariots of the Gods?* for their letters and suggestions. I want them to accept this book as the response to their encouragement.

I should like to thank everyone who helped me to write this new book. I wrote it during my imprisonment on remand in the Remand Prison of the Canton of Graubünden in Chur.

ERICH VON DÄNIKEN

*RETURN TO THE STARS*

# 1

# INTERSTELLAR SPACE FLIGHT IS POSSIBLE

When Thomas Edison invented his carbon filament lamp in 1879, gas shares fell overnight. In England, Parliament set up a Committee of Inquiry to examine the future possibilities of the new method of lighting. Sir William Preece, Postmaster-General and Chairman of the Committee, told the House of Commons that it had reached the conclusion that electric light in the home was fanciful and absurd.

Today electric lights burn in every house in the civilised world.

Obsessed by man's age-old dream of being able to fly, Leonardo da Vinci spent years secretly working on the construction of flying machines that were amazingly like the prototype of the modern helicopter, but he hid his sketches for fear of the Inquisition. When they were published in 1797, the reaction was unanimous that heavier-than-air machines could never leave the ground. Even at the beginning of this century the celebrated astronomer Simon Newcomb thought that a motive force powerful enough to enable flying machines to cover long distance was inconceivable.

Yet only a few decades later aeroplanes were carrying tremendous loads over land and sea.

Reviewing Professor Hermann Oberth's book *Rockets to Planetary Space* in 1924, the world-famous periodical

*Nature* commented that a space rocket project would probably only become practicable shortly before mankind became extinct. Even during the 1940's when the first rockets had already been launched from the earth's surface and flown hundreds of miles, doctors insisted that any kind of manned space travel was impossible because the human metabolism would be unable to stand the condition of weightlessness for several days on end.

Yet mankind has not died out and rockets are a familiar sight, and contrary to all predictions the human metabolism can obviously stand the condition of weightlessness.

What I am saying is that at some time or other the technical feasibility of every new idea vitally affecting the life of mankind was 'not proven'. Proof of its practicability was always preceded by the speculation of the so-called visionaries who were violently attacked, or what is often harder to stomach, laughed at condescendingly by their contemporaries.

I admit quite frankly that in this sense I am a visionary, too, but I do not live in splendid isolation with my speculations. My conviction that intelligences from other planets have visited the earth in the remote past is already under serious consideration by many scientists in both East and West.

For example, Professor Charles Hapgood told me during my stay in the USA that Albert Einstein, whom he had known personally, was in complete sympathy with the idea of a prehistoric visit by extraterrestrial intelligences.

In Moscow Professor Josef Samuilovich Shklovsky, one of the leading astrophysicists and radio-astronomers of our day, assured me that he was convinced that the earth had received a visit from the cosmos at least once.

The well-known space biologist Carl Sagan (USA) also does not exclude the possibility that 'the earth has been visited by representatives of an extraterrestrial civilisation at least once in the course of its history'.

And Professor Hermann Oberth, the father of the rocket, told me in these words: 'I consider a visit to our planet by an extraterrestrial race to be extremely probable.'

It is gratifying to know that under the pressure of successful space flights science is beginning to concern itself intensively with ideas that were absolutely taboo only decades ago. And I am convinced that with every rocket that shoots into the universe the traditional opposition to my theory about the 'gods' will get weaker and weaker.

Only ten years ago it was absurd to talk about the existence of another form of intelligent life in the universe. Today no one seriously doubts that extraterrestrial life exists in the cosmos. When eleven scientific experts separated after a secret conference at Green Bank (West Virginia) in 1961, they had agreed on a formula which calculated up to fifty million civilisations in our galaxy alone. Roger A. MacGowan, who holds an important NASA appointment in Redstone (Alabama), even arrives at a figure of 130 milliard possible cultures in the cosmos, basing his theory on the most recent developments in astronomy.

These estimates seem comparatively modest and cautious if it proves to be true that the 'key of life'—i.e. the generation of all life from the four bases adenine, cytosine, guanine and thymine—controls the whole cosmos. If this *is* the case, the universe must be literally teeming with life.

Crushed by the facts, people reluctantly admit today that space travel within our solar system is conceivable, but in the same breath they say that they consider interstellar space travel impossible because of the vast distances. Like a conjurer, they whip out of the hat the statement that because interstellar space travel will never be possible in the future, our earth cannot have been visited in the past by unknown intelligences, because they would have had to have been able to traverse interstellar space. And that's that!

But why should not interstellar space travel be possible? Working on the speeds that seem possible to us today, it is calculated that the journey to our nearest fixed star, Alpha Centauri, which is 4.3 light years away, would take eighty years—in other words no man could survive the flight there and back. Is this calculation correct? Admittedly, the average expectation of life today is about seventy years. The training of space pilots is complicated; even the most intelligent young man would not be able to pass his examinations as an astronaut before he was twenty. And if he was over sixty, he would hardly be sent on a space mission. That leaves only forty years as an active astronaut. It sounds quite logical to say that forty years are not long enough for an interstellar expedition.

But that is a fallacy. Even a simple example shows why and simultaneously demonstrates how hard it is for us to escape from traditional patterns of thought whenever we tackle projects for the future. Let us suppose that I am given an accurate calculation showing that it is impossible for a water bacteria to move from point A to point B during its lifespan, because the microbe can only move at speed $x$ and neither the current nor the slope of the watercourse can increase the speed $x$ by more than $y$ per cent at most. That sounds convincing, but there is a mental error in the calculation. The water bacteria can move from A to B in many different ways. For example, we can freeze it. Then the ice cube containing the bacteria travels from A to B in an aircraft. The ice is melted and the microbe has reached its goal. Yes, someone will retort, if you switch off its vital functions. But it seems to me a perfectly feasible, indeed highly practical, method of transporting the microbe—just as I think (which is why I gave this example) that we have reached the stage when we must replace the obsolete methods by new ones.

In spite of all the objections to it, there is nothing far-fetched about my forecast that in the not too distant future it will be possible to freeze astronauts, thaw them out again on a set date and restore the use of their functions. Professor Alan Sterling Parkes, member of the National Institute for Medical Research in London, supports the view that by the early 1970's medical science will already have perfected a method of preserving organs for transplants indefinitely at low temperatures.

Anyway, the whole has always been equal to the sum of its parts, which is why I am convinced that my forecast is right.

In all experiments on animals, one problem that constantly recurs is how to keep the brain cells alive, since they die rapidly without oxygen. The fact that research teams of the US Navy, the US Air Force, as well as firms such as General Electric and the Rand Corporation, are working full time on it shows how seriously the solution of this problem is taken. The first reports of success come from the Western Reserve School of Medicine in Cleveland, Ohio, where the brains of five rhesus monkeys were separated from their bodies and kept functioning for as much as eighteen hours. The separated brains reacted unhesitatingly to noises.

These experiments are basically connected with the idea of constructing a 'cyborg' (the abbreviation of 'cybernetic organism'). In a speech the German physicist and cyberneticist Herbert W. Franke put forward the sensational idea that in the decades to come space-ships would journey to unknown planets without astronauts aboard and search the universe for extraterrestrial intelligences. Space patrols without astronauts? Franke assumes that the electronic equipment would be operated by a brain separated from a human body. This 'solo' brain, kept in a liquid culture medium which would have to be constantly replenished with fresh blood, would be the control centre of the spaceship. Franke thinks that the brain of an unborn child would be the most suitable for programming, because, not being burdened with mental processes, it could be fed with the data and information necessary for the special tasks of space travel. This programmed brain would lack the consciousness that makes normal brains 'human'. Herbert W. Franke says: 'Stimulations, as we know them, would be alien to the cyborg. It would have no feelings. The human solo brain is promoted to ambassador of our planet.' Roger A. MacGowan also predicts a cyborg, half living being, half machine. In the view of this scientific authority the cyborg will ultimately reach the stage of a completely electronic 'being', whose functions are programmed in a solo brain and translated into orders by the latter.

The Frankfurt Jesuit Paul Overhage who enjoys considerable fame as a biologist, said about this fantastic project for the future: 'Its realisation can scarcely be doubted because the rapid progress of biotechnology is constantly making it easier to carry out experiments of this kind.'

During the last two decades molecular biology and biochemistry have advanced very rapidly and achieved results which have completely changed a great deal of medical science and practice. The ability to slow down the process of ageing or even interrupt it completely lies within our grasp, and even the fantastic construction of a cyborg has already been removed from the realm of pure imagination.

Naturally these projects create moral and ethical problems which will perhaps be harder to solve than the actual medico-technical problems. But all this will fade into insignificance if we keep our eye on the other highly probable possibility that one day space-ships will reach such incredible speeds that they can traverse cosmic distances even within the normal life span of astronauts.

The explanation of this technical phenomenon lies in the time dilation effect, which is already an accepted scientific fact.

We must realise that 'terrestrial years' are quite irrelevant as far as passengers on an interstellar space journey are concerned. In a space-ship travelling just below the speed of light, time 'creeps by' slowly in comparison with the time that rushes along on the launching planet. This can be accurately calculated by mathematical formulae. Incredible as it may seem, we do not have to take these calculations on trust; they have been proved.

We must free ourselves from *our* conception of time, i.e. terrestrial time. Time can be manipulated by speed and energy. Our space-travelling grandchildren will break the time barriers.

Those who doubt the technical possibility of interstellar space travel adduce an argument that deserves close examination. They say that even if rocket propulsion units were ultimately built to reach a speed of 93,000 miles per second or more, interstellar space travel would still be impossible because at such a speed the minutest cosmic particles that struck the exterior of the space-ship would have the destructive and penetrative power of a bomb. Undoubtedly that objection cannot be rejected lightly, but how long will it be valid? Scientists in the USA and the USSR are already engaged on the development of electromagnetic safety rings to divert the dangerous particles floating in space away from space-ships. These research projects have already met with some success.

The sceptics also say that a speed of more than 186,000 miles a second is just a utopian dream, because Einstein has proved that the speed of light is the absolute limit of velocity. Even this counter-argument is only valid if one starts from the premise that space-ships of the future will have to be launched with the energy of millions of gallons of fuel and carried into the universe with the same source of energy. Today radar sets operate with waves travelling at 186,000 miles per second. But, the reader will ask me, what connection have waves with the propulsion of spaceships of the future?

In their book *The Planet of Impossible Possibilities*, the two Frenchmen Louis Pauwels and Jacques Bergier describe the fantastic project of the Soviet scientist K. P. Stanyukovich, who is a member of the Commission for Interplanetary Communications of the USSR's Academy of Sciences. Stanyukovich plans a space sonde which will be propelled by anti-matter. Since a sonde travels faster the faster the particles on board it are emitted, the Moscow Professor and his team hit upon the idea of constructing a 'flying lamp' that worked by the emission of light instead of red-hot gases. The speeds that can be reached in this way are enormous. As Bergier tells us: 'The passengers in such a flying lamp would not notice anything unusual. Gravity inside the space-ship would be the same as on earth. They would feel that time was passing in the normal way, yet in a few years they would have reached the most distant stars. After twenty-one years (by their time), they would be in the heart of the Milky Way, which is 75,000 light years from the earth. In twenty-eight years they would reach the Andromeda Nebula, our nearest galaxy, which is 2,250,000 light years away.'

Professor Bergier, a world-famous scientist, emphasises that these calculations have nothing to do with science-fiction, because Stanyukovich has verified in his laboratory a formula that can be checked by anyone who knows how to use a table of logarithms. According to one of these calculations, only sixty-five years of cosmic time would pass for the crew of the 'flying lamp', while four and a half million years would go by on our planet!

Even in my wildest flights of fancy I cannot imagine the consequences of a development that is brewing in the dark mists of the future. In 1967 Gerald Feinberg, Professor of Theoretical Physics at Columbia University in New York, published his theory of tachyons in the scientific journal *Physical Review*. (Tachyon comes from the Greek word 'tachys' = fast.) His article was not just the ravings of a visionary; it described a piece of serious scientific research. A course of lectures on it has already been given at the Eidgenössischen Technischen Hochschule in Zurich.

The following is a brief description of the tachyon theory. According to Einstein's theory of relativity the mass of a body grows in relation to the increase in its velocity. A mass (= energy) that reached the speed of light would be infinitely large. Feinberg supplied mathematical proof that there was a counterpart to Einsteinian mass, namely particles that move infinitely fast, but become slower when they approach the speed of light. According to Feinberg, tachyons are a billion times faster than light, yet they cease to exist when they are reduced to the speed of light or below it.

Just as the theory of relativity (without which present-day physics and mathematics simply could not function) had only a mathematical proof for decades, tachyons are not yet demonstrable experimentally, but only mathematically. However, Feinberg is working on an experimental proof.

Believing in the future as I do, my fantasy runs away with me when I hear about research of this kind. Time and again during the last hundred years we have ultimately lived to see things that were considered impossible in the form of industrially manufactured products. So on this occasion I think I am entitled to enlarge upon an idea that, as I have said, is still in its infancy.

What might happen in the future?

If it became possible to capture tachyons or produce them

artificially, they could also be transformed into the propulsive energy for space sondes. Then, I assume, a space-ship would be brought up to the speed of light using a photon propulsion unit. As soon as it was reached, a computer would automatically switch on the tachyon propulsion unit. How fast would the space-ship travel then? A hundred, a thousand times faster than light? No one knows the answer today. Scientists suspect that once past the speed of light so-called Einsteinian space would be left behind and the space-ship hurled into an as yet undefined, superimposed space. But at this vital moment in the history of space travel the time factor would become almost meaningless.

I know of many fields of research in which the work going on is mainly devoted to the service of interstellar space travel. I have been in many laboratories and talked to many scientists. No one knows how many physicists, chemists, biologists, atomic physicists, parapsychologists, geneticists and engineers are working on the project that will enable man to fly back to the world of the stars—they are often lumped together, somewhat inaccurately, under the generic name of futurological projects.

It seems to me to be an under-evaluation of the human potential if, under the pressure of proof provided by technological advances, people admit the possibility of investigating cosmic space at some future date, but obstinately deny that the universe may contain intelligences who knew all about interstellar space travel thousands of years before we did and so could have visited our planet.

Because it has long been the custom to hammer into us as schoolchildren the presumptuous idea that man is the 'lord of creation', it is obviously a revolutionary and unpleasant thought that many thousands of years ago there were unknown intelligences who were superior to the lord of creation, but however disagreeable it is, we had better get used to it.

# 2

# ON THE TRACK OF LIFE

In my book *Chariots of the Gods?* I put forward the speculative idea that 'God' had created man *in his own image* by means of an artificial mutation. I voiced the suspicion that *homo sapiens* became separated from the ape tribe by a deliberately planned mutation. I have been attacked for making this assertion.

Because the tracing back of the origin and development of man has so far been limited entirely to our planet, my theory that extraterrestrial beings could have had a hand in the process is a bold one. Yet if this idea was considered within the bounds of possibility, our beautiful family tree—apes scrambled down from the trees, mutated and became the ancestors of man—would be destroyed. Since Charles Darwin (1809–1882) put forward his theory of natural selection, all fossil finds, from the skeletons of primitive apes up to *homo sapiens*, have seemed to be convincing arguments in favour of his theory. When the teacher Johann Carl Fuhlrott (1804–1877) discovered some old bones in the Neanderthal near Düsseldorf and used them to reconstruct 'Neanderthal Man', who lived in the last interglacial age and at the beginning of the Würm glaciation, i.e. from about 120,000 to 80,000 years ago, he built up his theory of ape men on the basis of this find. It caused quite a stir in scholarly circles. Disconcerted religious opponents of Fuhlrott's theory put forward the rather unconvincing argument that there could not be a fossil man, because fossil men ought not to exist.

There are many other species in addition to Neanderthal man. The lower jaw of a kind of primate was found at El Fayum near Cairo. It was dated to the Oligocene Age, which was between thirty and forty million years ago. If that dating is correct, it would be proof that beings resembling man must have existed long before Neanderthal man. Fossil examples of hominids have also been found in England, Africa, Australia, Borneo and elsewhere.

What do these finds prove?

They prove that we cannot say anything definite, because nearly every new find throws doubt on the datings that have just been included in our text-books. In spite of the large number of finds, it is fair to say that they give very inadequate clues to the historical continuity of the origin and development of the human race. Certainly the track of racial development from hominids to *homo sapiens* can be followed back clearly for millions of years, but we cannot make nearly so definite a statement about the *origin of intelligence*. There are minimal indications from the remote past, but they do not add up to a whole. So far I have not been fortunate enough to hear an explanation of the origin of intelligence in man that is even tolerably convincing. There is a vast number of speculations and theories about how this 'miracle' is supposed to have happened. That is why I believe that my theory is equally entitled to a hearing.

In the course of the thousands of millions of years of general evolution, human intelligence seems to have appeared almost overnight. If we reckon in millions of years, we can say that this event must have taken place 'suddenly'. While still anthropoids, our ancestors created what we call human culture astonishingly quickly. But intelligence must have made a sudden appearance for this to happen. Several million years passed before anthropoids came into being through natural mutations, but after that the hominids underwent a lightning-like development. All of a sudden, tremendous advances appear about 40,000 years ago. The club was discovered as a weapon; the bow was invented for hunting; fire was used to serve man's own ends; stone wedges were used as tools; the first paintings appeared on the walls of caves. Yet 500,000 years lie between the first signs of a technical activity, pottery and the first finds in hominids' settlements. Loren Eiseley, Professor of Anthropology in the University of Pennsylvania, writes that man emerged from the animal world over a period of millions of years and only slowly assumed human features. 'But,' he goes on, 'there is one exception to this rule. To all appearances his brain ultimately underwent a rapid development and it was only then that man finally became distinguished from his other relatives.' Who was it that taught us to think?

Although I have great respect for the hard work done by anthropologists, I must frankly admit that I am not really interested in the prehistoric age that the eyeteeth of anthropoids or hominids can be proved to belong to by fossil finds. Nor do I think the date when the first *homo sapiens* used stone tools very important. To me it is as obvious that primitive man was the most intelligent being on our planet as it is logical that the gods chose *this particular being* for an artificial mutation. I am far more interested in when primitive man first introduced moral values such as loyalty, love and friendship into his communities. What influence were our ancestors under when they experienced this change? Who imparted the feeling of reverence? Who implanted the feeling of shame in connection with the sexual act?

Is there a plausible explanation of why savages suddenly clothed themselves? We are given vague hints about drastic changes or fluctuations in the climate. We are also told that the anthropoids wanted to adorn themselves. If that is the correct explanation, the gorillas, orangoutangs and chimpanzees living in the jungle could gradually have begun to wear trousers or use ornaments.

Why did the anthropoids suddenly begin to bury their fellows when they had only just outgrown an animal existence?

Who taught the savages to take the seeds of certain *wild* plants, pound them up, add water and bake an article of food from the resulting mush?

Why anthropoids, hominids and primitive men learnt nothing for millions of years and then suddenly primitive men learnt so much is a question that nags at me. Has too little attention been paid to this important question up to the present?

The field of research devoted to the explanation of the origin of mankind is interesting and very worthwhile.

Yet the question why, how and from what date man became intelligent seems to me to be at least as interesting.

Loren Eiseley writes: 'Today on the other hand we must assume that man only emerged quite recently, because he appeared so explosively. We have every reason to believe that, without prejudice to the forces that must have shared in the training of the human brain, a stubborn and long drawn out battle for existence between several human groups could never have produced such high mental faculties as we find today among all peoples on the earth. Something else, some other educational factor, must have escaped the attention of the evolutionary theoreticians.' That is precisely what I suspect. There is a decisive factor that has not been taken into account in all the theorising on the subject. I doubt if we shall be able to supply the missing link without investigating the theory of visits to our planet by extraterrestrial intelligences and checking whether these beings should not be held responsible for an artificial transformation of hereditary factors, for a manipulation of the genetic code and for the sudden appearance of intelligence. I have something to say along these lines that strengthens my theory that man is a creation of extraterrestrial 'gods'.

In 1847 Justus von Liebig wrote in the 23rd of his *Chemical Letters*: 'Anyone who has observed ammonium carbonate, phosphate of lime or potash will obviously consider it quite impossible that an organic germ capable of reproduction and higher development can ever be formed from these materials by the action of heat, electricity or other natural forces...' The great chemist also claimed that only a dilettante could imagine that life had originated from dead matter. Today we know that this did happen.

Modern research assumes that the first life on earth originated one and a half milliard years ago. Professor Hans Vogel writes: 'In those days the barren land and the vast primordial ocean were enveloped in an atmosphere that was still without oxygen. Methane, hydrogen, ammonia, steam and perhaps acetylene and cyanide of hydrogen as well, formed a covering around the earth, which was still devoid of life. That is the kind of environment in which the first life must have originated.'

In their efforts to get on the track of the origin of life, scientists tried to make organic matter originate from inorganic matter in the conditions of the primitive atmosphere.

The American Nobel Prize winner Professor Harold Clayton Urey surmised that the primitive atmosphere had a composition far more susceptible to penetration by ultraviolet rays than our own. So he encouraged his colleague, Dr Stanley Miller, to check experimentally whether the amino acids necessary for the existence of all life would be formed in a primitive atmosphere created in a retort and subjected to radiation. Stanley Miller began his experiments in 1953.

He built a glass container in which he produced an artificial primitive atmosphere made up of ammonia, hydrogen, methane and steam. So that the experiment might take place in sterile conditions he had the Miller apparatus, as it is now known in scientific literature, heated to a temperature of 180° Centigrade for eighteen hours. In the upper half of the glass sphere he fixed two electrodes, between which electrical discharges were constantly flying. In this way, using a high frequency current of 60,000 volts, a permanent miniature storm was produced in the primitive atmosphere. In a smaller glass sphere sterile water was heated and its steam was conducted via a tube to the large sphere containing the primitive atmosphere. The cooled off matter flowed back into the sphere containing sterile water, to be reheated there and so climb up again to the sphere containing the primitive atmosphere. In this way Miller had created in his laboratory a cycle of the kind that had gone on on earth in the beginning of time. This experiment continued for a whole week without stopping.

What came out of the primitive atmosphere that was subjected to the steady lightning flashes of the miniature storm? The 'primitive soup' Miller had cooked up contained asparagine, alanine and glycine—in other words amino acids necessary for the building up of biological systems. In Miller's experiment complicated organic combinations had originated from inorganic matter.

During the years that followed countless experiments along the same lines were carried out under different conditions. Finally twelve amino acids were produced and now no one doubts that the amino acids necessary for life can originate from the primitive atmosphere.

Other scientists used nitrogen instead of ammonia, formaldehyde or even carbon dioxide instead of methane. Miller's electrical discharges were replaced by supersonic

30

waves or ordinary light waves concentrated into one beam. There was no change in the results! All the different primitive atmospheres, none of which contained a trace of

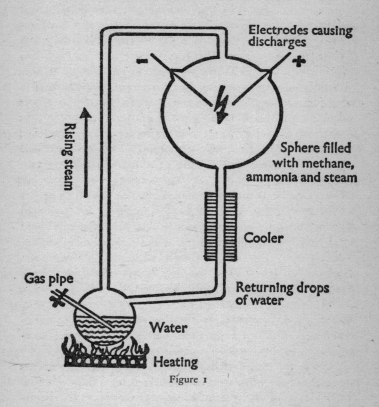

Electrodes causing discharges

−     +

Rising steam

Sphere filled with methane, ammonia and steam

Cooler

Gas pipe

Returning drops of water

Water

Heating

Figure 1

organic life, produced amino acids and nitrogen-free organic carbonic acids every time. In some experiments the specially treated atmosphere even produced sugar.

What are we to make of this phenomenon?

Ever since man has been able to think, he has tried to

evaluate everything around him in terms of polarity: light versus dark, heat versus cold, life against death. The habit of describing all living matter as 'organic' and all inanimate matter as 'inorganic' also falls into the broad field of this evaluation by polarities. But just as there are many intermediate stages between all extreme designations, it has long been impossible to draw a rigid boundary line between organic and inorganic chemistry.

When our planet began to cool off, what we call the 'primitive atmosphere' was formed from the light matter whose gas molecules were swirling about in confusion. It consisted mainly of those ingredients from which Miller brewed up his primitive soup in his laboratory experiments. Owing to the original high temperatures of the earth and its weak gravity, light gases such as helium and free hydrogen were lost in the cosmos, while the heavy gas molecules such as nitrogen, oxygen, carbon dioxide and also the heavier 'noble' gas atoms were retained. Hydrogen in its free state, its elementary form, is now virtually non-existent in our atmosphere; it is only found in chemical combinations. For example, two atoms of hydrogen together with one atom of oxygen form a molecule of the essential compound water (chemical symbol: $H_2O$).

The cycle got under way. Water evaporated and rose with the warm radiation from the earth in clouds of vapour which cooled off at great heights and poured down as rain. This primitive rain freed many kinds of inorganic matter from the hot stone crust and swept them into the ocean. Inorganic compounds such as ammonia and cyanide of hydrogen from the atmosphere also dissolved in the primitive ocean and took part in chemical reactions. For millions of years the earth's atmosphere grew richer in oxygen.

This development took place slowly. Today science is unanimous in saying that the transformatin of the original atmosphere into our oxidising atmosphere took about 1·2 milliard years. At the beginning of this development was the primitive soup, which, with its numerous forms of matter in solution, was a first-class culture medium for the first primitive forms of life.

It is said that life is always connected with an organism, in the simplest case with the organism's cell. The fact that an organism lives is proved by its metabolism, and also shows in its development. Functions constitute life. Are all these currently accepted criteria necessarily correct? If they are, a virus does not live. A virus itself undergoes no change of matter and energy; it does not eat nor does it excrete. It only multiplies inside foreign cells by reproduction. It is a parasite.

What then is life?

Shall we ever be able to define it?

If we follow the path of the origin of life in its main stages, the vital question is: whence the first living cell? Theodor Schwann (1810–1882) and Matthias Schleiden (1804–1881) carried out the fundamental research. Schwann proved that animals and plants are made up of cells; Schleiden recognised the importance of the nucleus. Then the Augustine prior Gregor Johann Mendel (1822–1884), who taught natural history and physics at Brün, made his cross-breeding experiments with peas and beans. This progressive priest discovered three laws of heredity with his patient experiments and became the founder of the science of heredity. Today his laws are unanimously accepted as governing men, animals and plants.

About the middle of the nineteenth century it was proved that the cell is the carrier of all vital functions. This proof became the basis for all the big biological discoveries. Now new techniques (Röntgenology, electrophoresis, ultramicroscopy, phase-contrast microscopy, etc.) enable us to examine cells and nuclei.

We suspect that the information centres for the storing

and transmission of hereditary factors are in the cells and the nuclei.

Research in this field, which is still comparatively recent, found out that every organism has a specific number of chromosomes, which have their own specific shape. Chromosomes are the carriers of hereditary factors. For example, the cells of the human body have 23 pairs of chromosomes = 46 chromosomes, a bee's cells have 8 pairs = 16 chromosomes, a sheep's cells 27 pairs = 54 chromosomes.

The protein molecules of the cells consist of chains of amino acids. Given this piece of scientific information, we were faced with the new question of how living cells originated from amino acids.

In connection with the only partially solved problem of how protein could come into being before there were living cells, Rutherford Platt describes the theory held by Dr George Wald of Harvard University. Wald assumed that under certain natural conditions amino acids must give the answer. Dr S. W. Fox of the Institute for Molecular Evolution, Miami, tested this idea by drying out solutions of amino acids. Fox and his collaborators observed that the amino acids formed long thread-like sub-microscopic structures. They had formed chain compounds containing hundreds of amino molecules. Dr Fox called them 'protenoids', i.e. protein-like matter.

Following the investigations of Professors J. Oro and A. P. Kimball, the chemists Dr Matthews and Dr Moser succeeded in producing protein matter from poisonous prussic acid and water in 1961. Three scientists from the Salk Institute, Robert Sanchez, James Ferris and Leslie Orgel, managed to produce synthetically the nucleic acids essential for metabolism and reproduction—those combinations of nucleic bases, carbohydrates and phosphoric acid occurring in the nuclei.

After our brief canter through chemistry and biology, the main thing for the reader to have grasped is that the construction of a living organism is a chemical process. 'Life' can be produced in laboratories. But what connection have nucleic acids with life?

Nucleic acids determine the complicated process of heredity. The sequence of four bases—adenine, guanine, cytosine and thymine—gives the genetic code for all forms of life. Once this discovery was made, chemistry was able to remove a great deal of the mystery surrounding life.

There are two groups of nucleic acids whose names have become familiar to every assiduous newspaper reader: RNA (ribosenucleic acid) and DNA (deoxyribosenucleic acid). Both RNA and DNA are necessary for the synthesis of protein in the cells. It is a fact that the proteins of all organisms examined so far are built up of about twenty amino acids and that the sequence, or arrangement, of amino acids in a protein molecule is determined by the sequence of the four bases in the DNA (=the genetic code).

But even if we know the structure of the genetic code, we are still a long way from being able to read the information stored in a chromosome. Nevertheless the thought that twenty amino acids are the bearers of all life and that their own arrangement in protein molecules is laid down in the genetic code is earth-shaking. In his book *The Biological Timebomb* George Rattray Taylor quotes the views of the Nobel prize-winners Dr Max Perutz and Professor Marshall W. Nierenberg on the tremendous possibilities that lie ahead.

Dr Max Perutz says: 'There are about one hundred million pairs of nucleotide bases distributed among forty-six chromosomes in a single human cell. How could we erase a specific gene from one particular chromosome, or add one to it, or repair a single pair of nucleotides? It hardly seems practicable to me.'

Professor Marshall W. Nierenberg, who played a vital role in the discovery of the genetic code, has a quite different opinion. 'I have no doubt that the difficulties can be overcome one day. The only question is when. I imagine that we shall succeed in programming cells with synthetic genetic information within the next twenty-five years.'

Lastly Joshua Lederberg, Professor of Genetics at Stanford University in California, is convinced that we shall be able to manipulate all our hereditary factors within the next ten or twenty years.

At all events, we have now realised that an insight into hereditary factors and their transformation is possible. And since we human beings know this, there is really no reason why an extraterrestrial intelligence that is familiar with space travel and consequently thousands of years ahead of us scientifically should not know it too.

In their book *You will live to see it*, the physicist and mathematician Herman Kahn, Director of the Hudson Institute of New York, and Anthony J. Wiener, adviser to the American government and also a member of the Hudson Institute, quote the *Washington Post* for 31.10.66, which described the possible results of manipulating the genetic code:

'Within ten or fifteen years a housewife will be able to go into a special store, look through a selection of packets like seed packets and choose her child by the label. Each packet will contain a one-day-old frozen embryo and on the label the buyer will be able to read the colour of hair and eyes, the size of body and IQ she can expect. There will also be a guarantee that the embryo has no hereditary defects. The woman will take the embryo of her choice to her doctor and have him implant it. Then it will grow in her body for nine months just like her own child.'

Such forecasts of the future are possible because the DNA contains genetic information for building the cells, as well as all the other hereditary factors. The DNA is a perfect 'punched card' for the structure of all life. For it not only codifies the twenty amino acids, but also announces the beginning and end of a protein chain with 'start' and 'stop', like a punched card prepared for a modern accounting machine. And just as the central unit of an electronic calculator contains a control bit whose job is to check all the calculating operations, there is a constant check on the functioning of the DNA chains in the cells.

James D. Watson, who investigated the structure of the DNA molecule so brilliantly at the age of twenty-four, has described the course of his work in his book *The Double Helix*. For the 900-word article in *Nature*, in which Watson described the bizarre spiral staircase shape of the DNA molecule, he and his fellow-workers received the Nobel Prize in 1962. Yet his book came within an ace of not being published. The board of the Harvard University Press opposed his frank way of describing things. They were afraid that the myth of ascetic scientific research might be destroyed by Watson's uninhibited narrative. For he says quite bluntly that he owed his success mainly to the preliminary studies and mistakes of his colleagues.

A spectacular event took place in America in December 1967. President Lyndon B. Johnson personally announced a great scientific achievement at a press conference in these words:

'This will be one of the most interesting articles you have ever read. An awe-inspiring achievement! It opens the door to new discoveries, to the disclosure of the fundamental secrets of life.'

What kind of event was so important that the President of the United States took such an interest in it?

Scientists of Stanford University at Palo Alto, California, had succeeded in synthesising the biologically active nucleus of a virus. Following the genetic pattern of a type of virus called Phi X 174, they had constructed from nucleotides one of the giant DNA molecules that control all vital processes. The Stanford University scientists put artificial virus nuclei into host cells. The artificial viruses developed just like natural ones. Parasites that they are, they bullied the host cells into producing millions of new viruses following the pattern of Phi X 174. Just as happens in an organism attacked by a virus infection, the artificial viruses burst through the host cells once they had used up their vital energy.

Obeying the orders given by the DNA molecule, the cells produced millionfold combinations of protein molecules from amino acids. Each new combination corresponded exactly to the programmed sample. The Californian scientists calculated that only one 'genetic misprint' occurred in the creation of one hundred million new cells.

Barely fifteen years after the explanation of the DNA structure by Watson, Crick and Wilkins, an important scientific discovery was made. The Nobel prizewinner Professor Kornberg and his colleagues succeeded in deciphering thousands of combinations of the genetic code for the virus Phi X 174. They had produced life in the laboratory in California.

Many readers will ask what these biochemical digressions have to do with the theme of my book? I have followed these investigations with keen interest ever since they were first reported. Why?

The conclusions reached convinced me that they had logical consequence, which Sir Bernard Lovell, founder and Director of the radiotelescope station at Jodrell Bank, formulated as follows:

'In the last two years it seems that the discussion of the question whether life exists outside our earth has become both serious and important. The seriousness of the discussion is a consequence of current scientific views, according to which the development of our solar system and of organic life on earth is probably not a unique case.'

In the summer of 1969 *Physical Review Letters* announced that American scientists, using the radiotelescope at Green Bank, West Virginia, had proved the existence of formaldehyde in gas and dust clouds in the universe. Formaldehyde, which is used in chemistry as a preservative and disinfectant, among other things, is a colourless, unpleasantly acrid-smelling gas. This, the most complicated chemical compound in space to date, which was ascertained from fifteen out of twenty-three sources of radiation by the American scientists, extends the list of primitive substances which are accepted as building stones of life by way of amino acids. This news provides new fuel for the suspicion that life is present in the cosmos.

But if there is life on other planets, I think it likely that unknown cosmonauts brought with them to our earth branches of knowledge of the kind we are now acquiring and that they made our ancestors intelligent by manipulation of the genetic code.

In the earlier account of the creation in the Bible we read (Genesis, v, 1–2):

'In the day that God created man, in the likeness of God made he him.

'Male and female created he them; and blessed them, and called their name Adam, in the day when they were created.'

According to my speculations, this could only have taken place by an artificial mutation of primitive man's genetic code by unknown intelligences. In that way the new men would have received their faculties suddenly—consciousness, memory, intelligence, a feeling for handicrafts and technology.

In the later biblical story of the creation (Genesis, ii, 21–23), we find a different version of the origin of woman:

'And the LORD GOD caused a deep sleep to fall upon Adam, and he slept: and he took one of his ribs, and closed up the flesh instead thereof;

'And the rib, which the LORD GOD had taken from man, made he a woman, and brought her unto the man.

'And Adam said, This is now bone of my bones (!), and flesh of my flesh (!): she shall be called Woman, because she was taken out of a Man.'

It is quite possible that woman was created from man, but Eve can hardly have blossomed forth in her naked beauty from the narrow rib of the male thorax by a conjuring trick—after a surgical intervention? Perhaps she originated with the help of a male sperm cell. But as according to the biblical Genesis there was no female human being in Paradise who could have received the seed, Eve must have been produced in a retort. Now a number of cave drawings showing objects like retorts in the vicinity of primitive man have been preserved. Could foreign intelligences with a highly developed science and knowing about the immune biological reactions of bones have used Adam's marrow as a cell culture and brought the sperm to development in it? The comparatively easily accessible human rib would obviously have been the most suitable container for this biologically possible act of creation. That is a speculation, but one that is practicable in the present state of scientific knowledge.

As Eve was allotted as Adam's mate very suddenly in the Bible, there should have been a sudden spate of drawings of female beings on cave walls or Stone Age bones following the artificial creation of woman I have described. In fact there is plenty of confirmation of such a surmisal. The so-called 'mother goddesses' make their first appearance in the early Stone Age. Female Stone Age figures are found at La Gravette (France), Cukurca (southern Turkey), Laussel (France), Lespugne (France), Kostyenki (Ukraine), Willendorf (Austria), Petersfels (Germany) and elsewhere.

All these female figures are flatteringly described as 'Venuses'. In nearly all of them the artists have taken pains to emphasise the sexual organs and symptoms of pregnancy. Archaeology dates these early Stone Age figures to the Gravettian. We do not know what purpose they could have served nor why they appear for the first time in the Early Stone Age. It is conceivable that *homo sapiens* originated in different parts of the world in two different ways: by the planned mutation of the genetic code of hominids and by the artificial production of a female being and her cultivation in a retort.

Nevertheless, the 'new' men later mated with animals again. These lapses must be attributed to the old Adam, because he alone remembered coupling with apelike animals. After the artificial mutation it should have been natural for the new human beings to mate with each other. Henceforth every reversion to the former habit of mating with animals that led to reproduction was a step backwards. Can this backsliding have been the Fall of Man? And was it at the same time *original sin* against the building up of the new kind of cells?

A few thousand years later the 'gods' corrected this Fall. (I shall have more to say about this.) They destroyed the hybrid animal-men, separated a well-preserved group of new men and implanted new genetic material in them by a second artificial mutation.

Palaeoanthropologists are puzzled by the sudden, breathtakingly fast separation of the neanthropids, the group of *homo sapiens* to which we belong, from the family of prehominids, who were still ape-like in form. So far the process has been provisionally explained by a spontaneous mutation.

If we adopt the preanthropological datings marking the essential changes in our first ancestors for our theory of a planned artificial mutation by unknown intelligences, then the first artificial mutation with the genetic code used by the 'gods' must have taken place between 40,000 and 20,000 B.C. And the second artificial mutation would have occurred in more recent times, between 7,000 and 3,500 B.C.

Assuming these datings are correct, the 'gods' first visit would probably have taken place in the age which yielded the first drawings and figurines of women.

Anthropological scholars shy away from datings that go so far back. But surely the time dilation effect, which is unreservedly accepted by science, must have been valid in all ages?

Time dilation is a known quantity for all planned interstellar space travel projects, both now and in the future. The fact that it was governed by a law was first 'discovered' in our own day, but just because it is governed by a law it obviously held good at all times, and would have applied to the 'gods' who could have visited the earth in their spaceships travelling just below the speed of light.

Surely the time has come for anthropologists to take this scientifically verified phenomenon into account?

If they did, many of the questions about how our forefathers originated and became intelligent would be answered instantaneously.

No eternities have passed for the 'gods' since their visit to earth! If they had paid a visit to our planet thousands of terrestrial years ago, only a few decades would have passed for the crew of the space-ship.

Anyone who accepts time dilation as applying to the visit of the unknown astronauts will understand at once that the same 'gods' who developed woman from *homo sapiens* could also have given Moses the highly technical instructions for building the Ark of the Covenant.

I know that it is difficult to grasp and yet it could be true. I must explain once again that this is not necessarily all speculation. For a long time astronomy has been working successfully with these remarkable time shifts. The only thing that matters now is for archaeologists and pre-anthropologists to accept them, too.

## 3

# A 'SUNDAY' ARCHAEOLOGIST ASKS QUESTIONS

The 'Sunday' archaeologist has the great advantage of being able to give his imagination free rein and ask the specialists disconcerting questions. Naturally I make the most of this advantage to shake the platform on which many prehistoric findings are built up and which is supposed to be sacrosanct. Amateur investigators are embarrassingly industrious. They are assiduous collectors, readers and travellers, because they like to find the best ammunition for their theories in the hope that they will ultimately hit the bull's eye.

The Research Institute for Electro-Acoustics in Marseilles moved into a new building in the spring of 1964. A few days after the move several of Professor Vladimir Gavreau's fellow-workers began to complain of headaches, nausea and itching. Some of them were so badly affected that they trembled like aspen leaves. In an Institute devoted to electro-acoustical problems it seemed likely that some uncontrolled radiations in the laboratory were causing the mischief. Using hypersensitive measuring apparatus the scientists covered the building from top to bottom in an attempt to find out the cause of their colleagues' unfortunate condition. Find it they did. However, it was not the radiation of uncontrolled electrical frequencies. It was low frequency waves which had escaped through a ventilator and subjected the whole building to subsonic vibrations.

By one of those lucky coincidences which have so often helped research, Professor Gavreau had specialised in the investigation of sound waves for twenty years.

After the incident he said to himself that it ought to be possible to produce experimentally and deliberately what the ventilator had achieved unintentionally. So he and his colleagues built the first sound gun in the world in the Research Institute for Electro-Acoustics in Marseilles. Sixty-one tubes in a chessboard pattern were fixed to a grille. Then compressed air was blown through them steadily until a note of 196 hertz was given off. The result was devastating. Cracks formed in the walls of the new building; the stomachs and intestines of the laboratory workers began to vibrate painfully. The apparatus had to be switched off at once.

Professor Gavreau wanted to follow up this experiment, but first he had a protective device made for the sound gun's crew. Then he built a genuine 'death trumpet' which developed 2,000 watts and sent out sound waves of 37 hertz. This apparatus could not be tested at full strength in Marseilles because it would have sent buildings crashing to the ground over a radius of several miles. At present a 'death trumpet' seventy-five feet long is in the course of construction. It is expected to produce sound waves with the death-dealing frequency of 3.5 hertz.

Quite apart from the frightening vision such a 'death trumpet' conjures up for the future, it reminds us of an event in antiquity.

After the chosen people had crossed the River Jordan without getting their feet wet and besieged the town of Jericho with its twenty-one foot thick defence walls, the priests were given complicated instructions about marching round the city and blowing their 'trumpets'. The event is described in Joshua (vi, 20) as follows:

'. . . and it came to pass, when the people heard the sound of the trumpet, and the people shouted with a great shout, that the wall fell down flat, so that the people went into the city, every man straight before him, and they took the city.'

Neither the full blast of the priestly lungs nor a fanfare of many thousand trumpets could blow down walls twenty-one feet thick. But we know today that sound waves with deadly low hertz frequencies would have been perfectly capable of bringing down the walls of Jericho.

When Dr Mottier, an archaeologist at Berne University, and I were taking part in a debate on the Swiss Radio, she told me that giants had never existed because there were no fossil finds from which we could infer the existence of such a race.

However, Dr Lovis Burkhalter, former French delegate to the Prehistoric Society, holds a quite different opinion. In 1950, he wrote in the *Revue du Musée de Beyrouth*: 'I want to make it clear that the existence of gigantic men in the Acheulian age must be considered a scientifically proven fact.'

Which side is right? At all events, tools of abnormal size have been found that could not have been handled by men of normal stature.

Archaeologists excavated flint implements weighing nearly $8\frac{1}{2}$ lb near Sasnych (four miles from Safita in Syria). The flint tools found at Ain Fritissa (Eastern Morocco) are not to be sneezed at either. They were $12\frac{1}{2}$ ins long, $8\frac{1}{2}$ ins wide and weighed $9\frac{1}{4}$ lb. If we make a calculation based on the normal human stature and constitution, the beings who were able to handle these clumsy implements must have been about 12 ft tall.

Apart from finds of tools, at least three scientifically accepted finds point to the existence of giants in the past:

1. The Java giant.

2. The South China giant.
3. The South African (Transvaal) giant.

What race did they represent?

Were they lone-wolves?

Were they the wrongly programmed products of mutations?

Were they the direct descendants of gigantic cosmonauts from another world?

Were they especially intelligent beings with advanced technical know-how who had originated according to the genetic code?

The fossil finds give no conclusive answers to my questions. They are too meagre to form the stones for building up a proper genealogy. Will such a family tree ever be systematically investigated in some specially chosen region? Sensational discoveries are reported from time to time, but they nearly always turn out to be chance finds.

But documents—and we ought to take the old sources literally—clearly confirm the former existence of giants. Moses tells us in Genesis, vi, 4:

'There were giants in the earth in those days; and also after that, when the sons of God came in unto the daughters of men, and they bare children to them, the same became mighty men which were of old, men of renown.'

We get a graphic account in Numbers, xiii, 33:

'And there we saw the giants, the sons of Anak, which come of the giants: and we were in our own sight as grasshoppers, and so we were in their sight.'

Deuteronomy, iii, 11, even gives us details that allow us to make rough estimate of their physical stature:

'For only Og king of Bashan remained of the remnant of giants; behold, his bedstead was a bedstead of iron; is it not in Rabbath of the children of Ammon? nine cubits was the length thereof, and four cubits the breadth of it . . .'

(The Hebraic cubit is about 1 ft 9 ins!)

But the Pentateuch is not the only part of the Bible that speaks clearly and unequivocally about giants, the later books of the Old Testament also give descriptions of these supermen. Their authors lived at different times and in different places, so they could not have communicated with each other. Nor could the giants, as theologists sometimes claim, have been interpolated into the texts later in order to symbolise evil. If these apologists looked at the texts more closely, they could see that the giants always appear when performing perfectly practical tasks—waging war and single combat, for example—but never when moral concepts or moral behaviour are under discussion.

Besides, the documentation of giants is not confined to the Bible. The myths of the Mayas and Incas also recount that the first race created by the 'gods' before the Flood was a race of giants. They called two prominent giants Atlan (Atlas) and Theitani (Titan).

Just like our 'flying gods', giants haunt the world of sagas, legends and sacred books, but they are never put on the same footing as the gods in any of these sources. One serious handicap kept the giants on the earth; they could not fly. Only when a giant is clearly defined as the offspring of a god is he taken along on a heavenly journey. The giants usually appear as the gods' humble and obedient servants, who carry out their tasks, until they are finally described as stupid brutish creatures and their traces in literature are lost.

A scholar as serious as Professor Denis Saurat, Director of the Centre International d'Études Françaises in Nice, has made a serious study of giants. He definitely confirms that they once existed, and even those scholars who raise doubts sooner or later stumble over giants' graves, over menhirs, those vertical, roughly dressed stone blocks, which range up to 65 ft in height, over dolmens, burial chambers built of

48

Erich von Daniken in front of the Temple of the Inscriptions at Palenque (Mexico). Deep down in the interior of this pyramid is the tombstone of the god Kukulkan at the controls of his rocket.

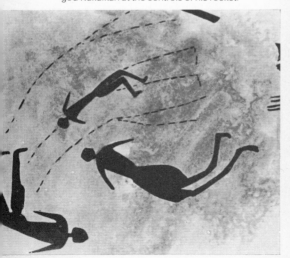

From time immemorial the idea of being able to float in the air has fascinated mankind – as in this Stone Age cave painting in the Libyan desert.

The figure in this rock painting from Ti-n-Tazarift in the Tassili mountains seems to be wearing a close-fitting space-suit with steering gear on his shoulders and antennae on his protective helmet.

A monument to the space-travelling gods? The most interesting thing on this stele from Santa Lucia Cotzumalhuapa (Guatemala) is the figure at the bottom right. It is dressed like a modern astronaut.

RIGHT: The Venus of Willendorf is the flattering name given to this limestone statuette with its faceless round head.

Man or hybrid of man and animal? This sculpture is known to archaeologists as the Man with the Catfish Head.

The significance of this female idol with four faces and a solar symbol, found at Lake Maracaibo (Venezuela), is unknown.

'Sunday archaeologist' on a journey of discovery through Mexico, autumn 1968.

This fresco from Sehar in the Tassili mountains shows, right, a figure 10 ft 8 ins high, surrounded by so-called men from Mars.

Erich von Daniken measuring the cyclopean walls above Sacsayhuaman (Peru).

This block of stone, the size of a four-storey house, has steps made with great accuracy. There is no credible explanation for it.

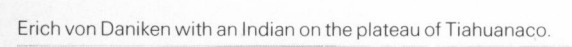

The picture shows the celebrated calendar of Sacsayhuaman. Has this monumental stone structure any relation to the ruins on the surrounding heights?

Erich von Daniken with an Indian on the plateau of Tiahuanaco.

onoliths that look as if they had been pre-cast like modern concrete. Thrones for giants? d the 'gods' destroy their base at Sacsayhuaman when they had carried out their ssion?

e rock seems to have been cut through as if it were butter. Who did it? When? How?

The famous 'water conduits' of
Tiahuanaco – (TOP LEFT) in this case
built into a temple wall of a recent
period for no apparent reason. The
'water conduits' have modern shapes
with smooth cross-sections, polished
inside and outside surfaces, and
accurate edges. But why place two
pipes side by side?

What were the clamps that once held this great block of stone together made of?

The angled section of a 'water conduit' from Tiahuanaco. Protective tubes for energy cables?

The Gate of the Sun at Tiahuanaco with its imposing frieze of figures.

massive stones, or other megalithic monuments, and last but not least over the impossibility of explaining technical achievements such as the working and transport of gigantic stone blocks. The number of gigantic architectonic edifices and the number of artistically dressed boulders we can still marvel at today can only be plausibly explained if we assume the primitive erectors of these works to have been giants or beings with techniques unknown to us.

Whenever I stand in front of a prehistoric monument on my travels, I always ask myself whether we ought to be satisfied with the previous explanations of its origin and purpose. Surely we ought to band together and have the courage to find out if novel and fantastic interpretations have any validity.

During my last journey in Peru in 1968 my friend Hans Neuner and I revisited the megalithic buildings above Sacsayhuaman (Falcon Rock), which is situated at a height of 3,450 to 3,780 ft near the limits of the former Inca fortress of Cuzco.

Tape measure and camera in hand, we approached these ruins, which are not ruins at all in the ordinary sense of the word. This is no heap of crumbled stone, remains of some historical building that have become unrecognisable. The rock labyrinth above Sacsayhuaman gives the impression of a super-edifice constructed with the last word in technical refinement. Anyone who has spent days in the thin air of this plateau clambering about among stone giants, caves and rock monstrosities, will find it hard to accept the explanation that all this was created ages ago by human hands using damp wooden wedges and crude stone mallets.

Here is only one of the examples we measured: a rectangle 7 ft 1 in high, 1 ft 2 ins wide and 2 ft 8 ins deep had been cut out of a granite block 36 ft high and 59 ft wide that appeared to have been torn from the cliff face. A first-class piece of work! There is nothing botched or crude about the way it has been extracted, there is no uneven or clumsy dressing. Even if we are prepared to admit the possibility that extremely skilled stonemasons managed to free the four lateral incisions of the colossus from the rock face after many years of work, we are still left with the riddle of how they freed the rear side of the rectangle. In those days the stonemasons certainly did not have cutting jibs of the kind used today when excavating the stone for underground submarine shelters. And presumably they did not possess the chemical knowledge to free the stone block from the rock face with the help of acids.

Or did they?

We climbed down into some caves in the rock that were 180 to 240 ft deep. As if shaken by some primaeval force, the caves' course has been interrupted and they are partially destroyed or telescoped together. Large sections of the ceilings and walls have been preserved. They are so perfect that they could compete with any present-day piece of pre-cast concrete. Nothing has been joined together; there are no parts held together by a binding agent. The whole thing looks as if it had come from one casting-mould. The edges are cut at right-angles and are knife sharp. Eight inch wide granite ledges lie stepwise as neatly as if the wooden mould had been taken away yesterday.

We walked upright through galleries and chambers, waiting tensely for the surprise awaiting us around the next turning. I kept on thinking about the current archaeological explanations of these masterpieces of technology, but they did not convince me. It seems much more likely to me that superlatively built fortifications must have existed here above Sacsayhuaman. All these faultlessly dressed stone colossi could have formed part of a megalithic building complex. Presumably this lay-out could be excavated or reconstructed if systematic research was carried out on the site.

Naturally I have also asked myself whether there might not be conventional explanations for the 'ruins' above Sacsayhuaman.

Volcanic eruptions? There have been none for miles around.

Movements of the earth's crust? The last violent movement is supposed to have taken place about 200,000 years ago.

Earthquakes? They could hardly have caused the damage which leaves so much order recognisable among the disorder. To add a double question mark after all the questions, the dressed granite blocks show signs of vitrification of the kind that only appears as the result of tremendously high temperatures.

Freaks of nature? The granite blocks have accurately cut grooves and they have mortises as if they had been torn loose from the block next to them.

Neither the city archaeologist at Cuzco nor his colleagues in the museums of Lima could give me a satisfactory explanation of the structures we had examined. 'Pre-Inca,' they said, 'or perhaps the Tiahuanaco culture.'

Of course, there is nothing shameful about admitting one's ignorance. The fact remains that no one knows anything definite about the blocks we saw above Sacsayhuaman. Only one thing is certain. This great complex was built by a method unknown to us by beings unknown to us at an unknown date. It is also certain that it existed and had already been destroyed again before the famous Inca fortress of the Sun God was built.

This applies equally to Tiahuanaco on the Bolivian plateau.

I had studied many books on the subject and learnt extraordinary things about Tiahuanaco in the process, but everything in them was surpassed by what I saw with my own eyes. I had also read a lot about the remarkable 'water conduits' that were discovered at Tiahuanaco. During my last journey to the Bolivian plateau I singled them out for special study.

There I stood in Tiahuanaco, 15,000 ft above sea level, for the second time. I had paid too little attention to the 'water conduits' during my first brief visit, but this time I wanted to rectify the omission.

I found the first remarkable example of these half-pipes set in the wall of a reconstructed temple. It had been put there arbitrarily. Where it sat in the wall the half-pipe was quite pointless, except perhaps in a decorative sense, as if it was aimed at the tourists.

When I was able to examine the 'water conduits' in other places, I found that what I had read about them was true. They had a completely modern shape with smooth cross-sections, polished inner and outer surfaces and accurate edges. The half-pipes have grooves and corresponding protrusions that fit together. They can be joined like children's Lego pieces.

If I was staggered by the technical and mechanical perfection of these works that the archaeologists attribute to pre-Inca tribes, I was absolutely flabbergasted when I saw that the finds long classified as 'water conduits' existed in the form of double pipes. *One* conduit was masterpiece enough, but now there were double pipes made out of one piece of rock. What is more, double pipes with faultlessly executed right-angled sections.

But how can anyone explain the fact that only the upper parts of the pipes have been found?

For the upper parts of 'water conduits' can be dispensed with, but not the lower pipes.

Did these stone pipes serve as water conduits at all?

Perhaps there is a quite different explanation, fanciful though it may seem.

Traditional legends and existing stone drawings tell us

that the 'gods' met at Tiahuanaco, before man was even created. In the language of our space age that means that unknown astronauts constructed their first base on the Bolivian plateau. They had a highly developed technology at their disposal, just as we today use laser beams, vibrating milling tools and electric apparatus. Looked at from that point of view, is it not more likely that the 'water conduits' were protective pipes for energy cables between individual buildings of the complex?

Beings who were capable of making pipes like those at Tiahuanaco must have possessed outstanding technical skills. Beings with such a high level of intelligence would not have been so stupid as to make water conduits with double pipes when they could simply have bored *a larger hole in the same stone to let double the amount of water pass through* by a very much simpler process and with very much less work. Intelligent beings with such abilities would not have chosen a right-angled shape for the transport of water because they would have known that water and dirt would collect in the corners. And naturally these technicians would also have made the lower sections of piping, if it was really needed for the transport of water.

When the Spanish *conquistadores* asked the natives about the builders of Tiahuanaco in the 1530's, they could give no information about them. They referred the Spaniards to the sagas, according to which Tiahuanaco was the place where the gods had created men. I suspect that the same 'gods' also made the pipes and that they did not use them as water conduits.

Archaeologists and anthropologists do their best to give dates to all historical finds. Once a find is dated, it assumes its predestined place in the existing system and, of course, it is given a catalogue number.

So far the C-14 method is the most accurate used by science. When using it, scientists start from the premise that the radioactive isotope of carbon (C), with the atomic weight 14, is present in our atmosphere in constant amounts. This carbon isotope is absorbed by all plants, so that trees, roots, leaves and grasses contain it in amounts that remain the same. But all living organisms absorb vegetable matter in some form or other, so men and animals also contain C-14 in the same proportion. Now radioactive substances have a definite period of disintegration, provided no new radioactive substances are added. This period of disintegration in men and animals begins with death, in plants with harvesting or burning. The carbon isotope C-14 disintegrates at the rate of one half its amount in about 5,600 years. This means that 5,600 years after the death of an organism, only half the original amount of C-14 is detectable, after 11,200 years only a quarter, after 22,400 only an eighth and so on. Since the original amount of C-14 in the atmosphere is known, the C-14 content of fossilised organic matter can be found out by a complicated laboratory process. In relation to the constant C-14 content of the atmosphere the age of a bone or a piece of charcoal can then be determined.

If grasses and bushes on the edge of a motorway are cut and burnt, the ashes give a false age of many thousands of years. Why is that? Day after day the plants have absorbed large amounts of carbon petroleum, but that in turn comes from organic material that stopped absorbing C-14 from the atmosphere millions of years ago. Thus a tree cut down in an industrial district may be only fifty years old according to its annual rings, but examinations by the C-14 method would date the wood ash so far back that the fifty-year-old tree would have had to have been planted in very remote times.

I doubt the accuracy and consequently the dependability of this method. Measurements made so far start from the firm assumption that the proportion of a C-14 isotope in the atmosphere is and was always the same.

But who knows if that is true?

And what happens if this premise is based on an error? In my book *Chariots of the Gods?*, I referred to ancient texts which said that the gods were capable of producing tremendous heat of the kind that only nuclear explosions can generate, and also that they were familiar with radiation weapons. In the *Epic of Gilgamesh* Enkidu dies be-

Are the symbols on this stone from Mexico just ornamentation or are these motifs from a forgotten technological age?

cause he has been smitten by 'the poisonous breath of the heavenly beast'. The Mahabharata tells us how warriors hurled themselves into the water to wash themselves and their armour, because everything was covered 'with the death-dealing breath of the gods.'

Supposing that in both this case and that of the 'explosion' in the Siberian Taiga on the morning of 30 June, 1908, what had really taken place was an atomic explosion?

Whenever and wherever—including Hiroshima, and all the nuclear weapon tests on Bikini atoll, in the Soviet Union, the Sahara and China—radioactive substances are released, the balance of the radioactive isotope $C$-14 must be disturbed. So plants, animals and men had and have more $C$-14 in their cells than the amount of $C$-14 in the atmosphere constantly accepted as a reference figure. This theory cannot be disputed. If it is accepted, then the so-called exact scientific datings would be called in question. In our theory of visits by unknown astronauts we are dealing with such vast periods of time that 'small' miscalculations can easily creep in and such a 'small' error can very quickly amount to 20,000 years or more.

That is one reason which makes me sceptical about datings that go very far back. Let us take the case of Tiahuanaco. If cosmonauts did leave our planet from there after completing their mission, they certainly did not leave any fossil heirlooms behind for the anthropologists and archaeologists. They had modern equipment and did not use charcoal fires for heating, and they took their bones away with them. In other words, they left absolutely no datable traces behind them. Bones and charcoal remains that are found on the presumed landing places of the astronauts and then analysed and dated come from men who settled among the ruins of the gods' fortress thousands of years later. I think it is a mistake to assume that the bones that have been excavated came from the builders of

Tiahuanaco. I ask new questions because the old answers do not satisfy me.

Archaeology has only existed as a scientific discipline for 200 years. Its representatives scrupulously collect coins, clay tablets, fragments of utensils, shards of vessels, figures, drawings, bones and everything the earth yields up to the spade. They arrange the finds neatly in a system that only has a relative validity for about 3,500 years. Anything that lies further back remains hidden behind a veil of riddles and suppositions. No one knows and no one can explain what made our ancestors capable of outstanding technical and archaeological achievements. People say a compulsive longing for the 'gods', the desire to please the 'gods', to carry out the duties imposed on them by the 'gods'—all these were the driving forces behind the many wonderful buildings.

Longing for the 'gods'?

*Which* 'gods'?

Carrying out duties imposed by 'gods'?

*Which* 'gods' imposed the duties?

'Gods' have to perform wonders; they have to be able to do more than other beings. Invented 'gods', pure products of the imagination, would not have stayed long in the consciousness of mankind. Men would soon have forgotten them. That is why I hold the view that the 'gods' of whom we speak must have been real figures who were so clever and so mighty that they made a deep impression on our ancestors and dominated man's ideological and religious world for many centuries.

Then who was it who manifested himself to the primitive peoples?

We should not be afraid to give fanciful theories a fair hearing. For unfortunately what Heracleitus (ca. 500 B.C.) once said still holds good today: 'Because it is sometimes so unbelievable the truth escapes becoming known.'

There is an area of ruins on the mountainsides of Cajamarquilla, east of the Peruvian capital of Lima. Evidence of our human past to which scholars have not yet paid enough attention is being destroyed there daily by voracious bulldozers engaged in making roads.

This painting with a heavenly snake, priests sacrificing (?) and strange flying objects comes from a Peruvian ceramic vessel now in The Linden Museum, Stuttgart.

We trudged through this wilderness. We did not need anyone to draw our attention to curiosities, for we literally stumbled into them. In the roads there are hundreds of fox-holes, like the ones dug by the Viet Cong which we know about from magazines and television. Of course, I cannot assert that these fox-holes were once dug to protect the inhabitants from air-raids. I dare not say that, because it is a well-known fact that no air-raids took place before the twentieth century!

In section the fox-holes of Cajamarquilla have a diamenter of 23 ins and a depth of 5 ft 7 ins. I counted 209 holes (!) in a single street. They must have served a very important and practical purpose, otherwise why all the expenditure of effort?

How did the local inhabitants explain the hundreds of holes?

They told me that they had once been grain silos.

This explanation is not very convincing when one

58

examines the holes, which are hollowed out to human size. Naturally such holes could be filled with grain, but surely it would soon begin to germinate or spoil in the damp ground and moist heat? And how was the grain shovelled out of the narrow silos again?

In the absence of grain we filled one of the silos with sand. Then we tried to scoop it out again with our hands and shovels. The upper third did not take much effort, but from the middle downwards the work became exhausting. The last third was sheer torture. With head hanging downward, one leant into the hole, scooped up a handful of sand, raised oneself up and put it near the edge. But then we reached a depth when we could no longer lift our hands past our heads without the sand trickling out of them. We soon put our shovels away because the narrowness of the shaft no longer permitted any leverage. Finally we tied small pails to ropes and lowered them into the depths. If we tried to fill them with the shovels, half the contents tipped out. We tried all kinds of dodges. After working all day and using every device we could think of, we had emptied one silo until only six to eight inches of sand were left. That remnant is probably still there today.

Ever since I was told that the numerous fox-holes were 'grain silos', one question has been bothering me. Why did the peoples of Cajamarquilla take such tremendous pains over the excavation of such narrow holes?

Why didn't they instal one big family silo?

As Cajamarquilla was obviously a well organised urban community, the idea of one large and practical communal silo must have suggested itself.

After investigating on the spot, I am by no means convinced that the time-honoured explanation is valid, but, people say, they *must have been* grain silos.

# 4

# MANKIND'S STOREHOUSE OF MEMORY

Why is it that we often cannot remember names, addresses, telephone numbers and ideas even when we try our hardest? We 'sense' that what we are looking for is hidden somewhere in the grey cells of our brain and is only waiting to be rediscovered. Where has the memory of something we know perfectly well gone to? Why cannot we make use of our store of knowledge as and when we require?

Robert Thompson and James McConnell of Texas spent fifteen years tracking down the secret of memories and their whereabouts. Having carried out all kinds of experiments, they finally made flatworms from the family with the sonorous name *Dugesia Dorotocephala* the stars of an experiment that was to lead to fantastic results. These creatures are among the most primitive organisms that possess cerebral substance, but at the same time they have a complicated structure which can be completely regenerated by cell division. If one of these little worms is cut in pieces, each single piece renews itself and becomes a complete and absolutely intact flatworm.

Thompson and McConnell let their starlets crawl about in a plastic trough, to which they connected a weak electric circuit. In addition, they installed their desk lamp with a sixty watt bulb above the trough. As flatworms are very aphotic (averse to light), they curled up every time the lamp was switched on. After the two scientists had repeated this game of 'light on, light off' for a few hours, the worms no longer took any notice of the constant changes. They had realised that no deadly peril threatened them, that darkness followed light and vice-versa. Now Thompson and McConnell combined the light stimulus with a weak electric shock, which affected the creatures a second after the light went on. Whereas the worms had ignored the light stimulus before, now they curled up again when they felt the electric current.

The worms were allowed a break of two hours before they were put on the rack again. Then it turned out that they had not forgotten that they must expect an electric shock after the light came on. They curled up after it was switched on even if the expected shock did not come.

Next the two patient investigators cut the worms into small pieces and waited for a month until the parts had regenerated to complete worms. Then they were returned to the test trough and the desk lamp was switched on at irregular intervals. Thompson and McConnell made the astounding discovery that not only the heads which had regenerated a new tail, but also the tail pieces which had built up a new brain curled up for fear of the expected electric shock!

Had chemical processes taken place in the cells which had stored the 'old' memories and transmitted the past experiences to the newly formed cells?

That was exactly what had happened. When an 'unskilled' flatworm devours a 'skilled' fellow creature, he takes over the abilities his victim has acquired. Experiments in other laboratories confirmed that if the cells from an animal that had been taught certain skills were inserted into the body of another animal the same skills continued to function. For example, rats were taught to press a specially coloured key if they wanted to reach their food. When the animals had completely mastered their lesson they were killed, an extract was taken from their brains and injected into the abdominal cavities of untrained rats. After only a few hours the untrained rats were already using the same coloured key. Experiments with goldfish and rabbits confirmed the assumption that learned knowledge can be transmitted from one body to another by the transfer of certain cells.

Today it is an accepted scientific fact that memories are stored in memory molecules and that RNA and DNA molecules establish and transport memory contents. If these investigations were carried a step further, mankind might be able in the foreseeable future to stop knowledge and memory disappearing when a man dies, and to preserve and transmit the intellectual possessions he has acquired.

Shall we live to see hyper-intelligent dolphins, 'programmed' in underwater research, go to 'diving stations'?

Shall we see apes whose brains have been 'programmed' to handle road-making machines working in the streets?

These things may sound ridiculous today, but I think that the man who doubts their practicability is sticking his neck out more than the man who reckons with them as serious possibilities.

As yet there is no scientific proof that unknown intelligences knew how to carry out this kind of memory manipulation in the remote past. Nevertheless, famous scientists such as Shklovsky, Sagan and others do not exclude the probability that there are living beings on other planets who have advanced far beyond our stage of scientific research.

Once again the Old Testament gives food for thought. It tells of several prophets who were given books to eat by the 'gods'.

Ezekiel iii, 3, recounts such a book feast: 'And he said unto me, Son of man, cause thy belly to eat, and fill thy bowels with this roll (of a book) that I give thee. Then did I eat it. . . .'

So it is not surprising that prophets 'nourished' in this way knew more than anyone else and were more intelligent than their contemporaries!

Since the scientific discovery of the DNA double helix, we know that the nucleus of the gene contains all the information necessary for the construction of an organism. Punched cards are so familiar today that to simplify things I should like to call the building plan that is programmed in the nuclei 'punched cards governing life'.

These punched cards build life according to a very precise time schedule. Let us take our own species as an example. A ten-year-old boy and an eight-year-old girl are obviously adults in miniature, but they do not possess many of the attributes they will have one day as man and woman. Before they have grown up, the cells in their bodies will have divided a trillion times and with each cell division new building stages will be summoned from the punched card. Boys and girls begin to grow rapidly; pubic hair, facial hair and breasts begin to show. The punched cards never make a mistake; their holes determine the course of growing up at exactly the right times.

I should like to emphasise once again that this is a fact that applies to every organism. So on this solid scientific basis I should like to put up for discussion an idea that I personally find quite logical. Why should there not have been a comprehensive building plan for the whole of mankind—as for every individual being—since the remotest times?

Anthropological, archaeological and ethnological facts embolden me to add my theory to other hypotheses of the origin of mankind. I suspect that in the case of primitive man all the information, i.e. all the orders punched in the cards, was introduced from outside by a planned artificial mutation.

If we follow my theory back into the dark maze of mankind's prehistory, man is both the 'son of earth' and 'child of the gods'. Tremendous and fantastic consequences result from this dual descent.

Our ancestors experienced 'their' age, the primordial past, directly. They took it into their consciousness and their memory stored up every event. As each new generation came into being a part of this primitive memory was transferred to it. Simultaneously each generation added new holes to the existing ones in the punched cards. The cards were constantly enriched with new information. Even if some information was lost in the course of time or had stronger impulses superimposed on it, the sum of all information did not decrease. But now man houses not only the punched holes of his *own* memories, but also the programming of the *gods*, who were already travelling in space in Adam's day!

Between our present knowledge and the wealth of these memories there stands a barrier which only a few men manage to break through at fortunate moments. Sensitive men—poets, painters, musicians and scientists—sense the creative stimulation of this primordial memory and often struggle desperately to recapture the stored up information. The medicine man puts himself into a trance with drugs and monotonous rhythms so that he can break through the barrier to the primitive memory. I also believe that even behind the trendy behaviour of the psychedelic pioneers a primitive instinct is at work that drives them to seek access to the unconscious by using drugs and exacerbating music. But even if the door to a buried world is sometimes opened in individual cases, most people are incapable of communicating to others what they experienced in their exalted state.

For example, Aladdin's lamp is always quoted when people want to describe a fantastic apparatus or miraculous process. But not only do I take the prophets literally, I have also got into the habit of seeking something real behind the strange primitive memories of the men of antiquity, something real that may only await (re)-discovery by us today.

What was the special thing about Aladdin's lamp? The fact that it could materialise supernatural beings whenever young Aladdin rubbed it. Is it possible that he set a materialisation machine going when he rubbed it?

In the light of the present-day knowledge there is a possible explanation of the magic lamp. We know that atomic technology turns mass into energy and that physics turns energy into mass. A television picture is split up into hundreds of thousands of lines that are radiated over relay stations after they have been transformed into waves of energy.

Let us take a leap into the fantastic. A table—including the one I am sitting at now—consists of a countless number of juxtaposed atoms. If it were possible to split the table up into its atomic components, send it out in energy waves and reconstruct it in its original form at a given place, the transport of matter would be solved. Sheer fantasy? I admit that it is today, but in the future?

Perhaps the memory of the men of antiquity was still haunted by the remembrance of materialisations that had been seen in very remote times. Today steel is dipped in liquid nitrogen to harden it. To us a natural process that was discovered in modern times. But probably owing to a primitive memory this hardening process was already a reality in antiquity. At all events, it was practised with very crude methods. For case-hardening, the men of old plunged red-hot swords into the bodies of live prisoners! Yet how did they know that the human body is pumped full of organic nitrogen? How did they know the chemical effect?

Simply by experience?

How, I ask, did our ancestors get their advanced technology and their modern medicinal knowledge if not from unknown intelligences?

How do intelligent men and women come to believe that some audacious, way-out idea is empirically arrivable

This mysterious drawing from Tell Issaghen II. Sahara is thought by some to show a mummy being transported. The two top figures seem to be floating in space.

at step by step, that what is originally fantasy or vision will one day become reality?

I am firmly convinced that scientists are inspired by the driving desire to know as many things, to turn into reality at least as many memories, as were introduced in the memory of mankind by unknown intelligences in the remote past. For there must be a plausible reason why the cosmos has been the great goal of research throughout human history.

Surely all stages of technical development, every tiny advance and all the visionary ideas were only steps towards the great adventure—the reconquest of space?

Ideas which we still find confusing and disturbing have probably already been turned into reality on our planet at some time in the past.

It was during my study of Teilhard de Chardin (1881–1955), whose books have made a great impact on many people today, that I first came across the concept of cosmic primordial parts. Only later ages will realise what a decisive say this Jesuit has had in forming the twentieth-century world picture with his palaeontological and anthropological researches, in which he wanted to combine Catholic teaching about the creation with the findings of present-day natural science. In 1962, seven years after his death, it was decided, after a violent theological dispute, that Teilhard's views violated Catholic doctrine.

I know of no concept that expresses so clearly what is meant by the cosmic processes. The primordial part of matter is the atom. The atom is also the material primordial part in the cosmos. But there are other primordial parts, for example, time, consciousness and memory. In ways as yet unexplained all these primordial parts are related and connected with one another. Perhaps one day we shall track down other primordial parts, i.e. forces, which cannot be defined as classified either physically, chemically or in other scientific categories. Yet even though they cannot be defined or conceived of materially, they have an effect on the cosmic process. And as far as I am concerned the frontiers where all research will and must end lie in the cosmos.

I sincerely hope that my observations will set up new signposts leading eventually to convincing results. Two cases which Pauwels and Bergier mention in their book

*Breakthrough into the Third Millennium* are directly in line with my conviction that primitive memories await their discovery in the human consciousness. There is nothing occult or esoteric about either of them. The first concerns the Danish Nobel Prize winner Niels Bohr (1885–1962), who laid the foundations for present-day atomic theory. This world-famous physicist related how the idea of the atom model he had sought for many years occurred to him. He dreamt that he was sitting on a sun of burning gas. Planets rushed past him, hissing and spitting, and all the planets seemed to be connected by fine threads to the sun around which they revolved. Suddenly the gas solidified, sun and planets shrivelled up and became motionless. Niels Bohr said that he woke up at this moment. He realised at once that what he had seen in his dream was the atom model. In 1922 he won the Nobel Prize for his 'dream'.

The second case mentioned by Pauwels and Bergier also concerns two natural scientists who figured both as dreamers and men of action. An engineer of the Bell Telephone Company in the USA read reports of the bombing of London in 1940. They upset him badly. One autumn night he dreamt he was drawing the design of an apparatus that could train anti-aircraft guns on the previously worked out path of an aircraft and ensure that their shells would hit the aircraft at a specific point regardless of its speed. The next morning the Bell engineer made a sketch of what he had already drawn in his dream. He finally built a set in which radar was used for the first time. The celebrated American mathematician Norbert Wiener (1894–1964) was in charge of the project for manufacturing it commercially.

I believe that what these two brilliant natural scientists 'dreamed' already rested on the basis of their 'age-old' knowledge. In the beginning there is always an idea (or a dream) that has to be proved in practice. I think it quite likely that one day the molecular geneticists, who already know how the genetic code functions, will also find out how much—and even which—information was programmed on the punched cards of our life by unknown intelligences. It sounds fantastic, but one fine day we might even discover by which code word a specific piece of knowledge for a specific purpose can be summoned up from the primitive memory.

In my opinion cosmic memories penetrated more and more strongly into our consciousness in the course of man's evolution. They encouraged the birth of new ideas, which had already been realised in practice at the time of the visit by the 'gods'. At certain fortunate moments the barriers separating us from the primitive memory fall. Then the driving forces brought to light again by the stored up knowledge become active in us.

Is it only a coincidence that printing and clockmaking, that the car and the aeroplane, that the laws of gravity and the functioning of the genetic code, were invented and 'discovered' almost simultaneously at different times in different parts of the world?

Is it pure coincidence that the stimulating idea of visits to our planet by unknown intelligences has appeared simultaneously and been put forward in a great many books with completely different arguments and sources?

It is, of course, extremely convenient to dismiss ideas as coincidences when there seem to be no cut and dried explanation for them, but that is too easy a way out. Scientists, who have generally tried hard to find rules behind all processes, should be the last people summarily to reject new ideas—however fantastic they may seem at first—as unsuitable for serious research.

Today we know that the plan for the growth and death of every organism is coded in its nucleus. But why should there not be a master plan for the whole of mankind, a great all-embracing punched card on which prehuman and cosmic memories are registered? This premise would explain once and for all why world-shaking ideas, discoveries and inventions suddenly come into existence at some given point in time. The point in time is programmed in the punched card! The selector picks out the storage points in the card and summons up forgotten and subconscious material.

The hectic rush of everyday life leaves us no leisure for getting to know the unconscious. Driven by a constant flood of new and stimulating impressions, our senses never reach the storage points of the primitive memories. So I find it logical that the wonderful sight of memories of the past and a vision of the future appears particularly to monks in their cells, scientists in the seclusion of their laboratories, philosophers in the solitude of nature and men dying alone.

Since the remote past we have all lived in an evolutionary spiral that carries us irresistibly into the future, into a future which I am convinced has already been the past; not a human past, but the past of the 'gods', which is at work in us and will become the present one day. We are still waiting for definite scientific proof. But I *believe* in the power of those chosen spirits to whom a subtle selective mechanism is given that will one day release to them the information stored up in the dim past about realities that have existed. Until that happy day dawns, I support Teilhard de Chardin when he says; 'I believe in science, but has science ever taken the trouble to consider the world except from the outside of things?'

# 5

# THE SPHERE THE IDEAL SHAPE FOR SPACE-CRAFT

All the types of rocket in service today are pencil-shaped. Is that absolutely necessary? Surely there is constant proof that the pencil shape is neither necessary nor ideal in airless space? When a space-ship, which, unlike the multi-stage ròcket, is at least cone-shaped, flies to the nearby moon, it has to revolve repeatedly on its own axis. How involved and dangerous! We know from all the accounts of space flight that every change of course calls for a highly compli- cated steering manoeuvre. The ship's computer has to find out deviations from the flight path in thousandths of a second and equally quickly actuate the small steering jets for course correction. A single, minute steering mistake would have devastating consequences, as only limited amounts of propulsive material are carried and they would soon be used up. Then the steering jets would no longer be able to carry out the course corrections, the space-craft would be unable to return to the earth's atmosphere and it would shoot through the universe out of control until it burnt up.

Undoubtedly the rockets now in use have proved them- selves technically. For with the present rocket motors, which are still comparatively weak, only pointed flying objects offering small frictional surfaces can pierce the thick 'wall' of the earth's atmosphere. Yet sharp needles are not ideal for interstellar traffic.

The liberation of higher propulsive energy is the key that would lead to the manufacture of new types of space-craft. The time when technology will have as yet incredible energies as its disposal is no longer so far away. When that time comes, it could lead to pure photon propulsion units that reach a velocity close to the speed of light and can provide propulsion for an almost unlimited period.

Then we should no longer have to economise on every pound of payload, as we do today, when for every pound that a space-craft takes on a journey to the moon, an extra 2,590 lb of fuel is needed. Once that was the case, space-craft would soon be built in a very different shape.

Old texts and archaeological finds around the world have convinced me that the first space-craft that reached the earth many thousands of years ago were spherical, and I am sure that the space-craft of the future will (once again) be spherical.

I am no rocket designer, but there are a couple of reflections that we can all make and which seem completely convincing. A sphere has no 'forward' or 'aft', no 'above' or 'below', no 'right' or 'left'. It offers the same surface in every position and direction. So the sphere is the ideal shape for the cosmos, which also has no 'above' or 'below', no 'forward' or 'aft'.

Let us take a walk round a space sphere that still seems like a science-fiction dream today. But let's not skimp matters. Imagine a sphere with a diameter of 17,000 ft. This monster stands on sprung, retractable spider legs. Like an ocean liner, the interior is divided into decks of various sizes. Around the belly of the gigantic ball—at its equator— runs a massive ring housing the twenty or more propulsion units that can all be swivelled through 180°—a simple technical feat. When the countdown has reached zero, they will radiate concentrated light waves amplified a millionfold. If the cosmic sphere is to rise from the surface of the planet or one of the launching areas stationed in orbit, the propulsion units shoot their columns of light directly down on to the launching pad, giving the sphere a tremendous thrust. Once the sphere has reached the extra-gravitational field and is on its course to a fixed star, the propulsion units around its equator will only be fired now and then for course corrections. There is no risk of the sphere moving out of its flight path in a way that might endanger the crew because it can immediately adapt itself to any situation. Besides, something happens that will be very pleasant for the astronauts; the sphere begins to rotate of its own accord. In this way an artificial gravity is created in all the external rooms that decreases the state of weightlessness so much that conditions are almost the same as on earth. If one flies to the stars, one is still bound by one of the laws of the old earth!

It is important to realise that in this kind of space sphere course corrections in any direction are possible without danger. The propulsion units mounted on the steel girdle round the sphere permit lightning avoiding action or quick turns in any direction. Billiards players will easily catch on to the idea. If a right turn is needed, the sphere gets a light touch from a steering jet mounted on the left and vice-versa.

Spherical space-craft of the kind that may have traversed the galaxies millennia ago will be only minute particles in the infinity of the universe. Shooting along close to the speed of light, the astronauts will only sense this tempo as a slow soft floating away. Time will seem to stand still in their craft.

But what will happen in the 'timeless time' in the interior of the cosmic sphere? Well, once space stations of that size actually travel, a perfectly normal everyday routine will be followed on board. Robots will keep a check on the functioning of motors and machinery, computers will watch the course, astronauts will carry out scientific research in laboratories, think out still bolder projects, observe the stars and think about the exploitation of unknown planets. While the sphere covers millions of miles a minute, days will become weeks, weeks will become months and months become years for the crew. And in deep-freeze sarcophagi a reserve crew will await its biological reawakening when the sphere nears its goal.

But simultaneously on countless planets whole cultures will disappear, generations will die and new ones will be born, for time will rush by according to 'terrestrial' laws on our planet and other stars.

I won't expand on the excursion to Utopia. Science-fiction writers have described imaginary space-ships of the future only too often. My 'sphere story' is solely intended to prepare the reader's imagination for a perfectly serious idea. Supposing we examine the first legends of mankind's creation with this 'sphere story' in mind?

We learnt at school that in the beginning there were only heaven and earth and that the earth was deserted and barren. But out in the darkness, we were taught, shone a light and from this light came the word which gave the order for all life to begin.

Everything about the temporal unfolding of this genesis is absolutely logical. During the long cosmic journey through the universe there was obviously no light; all was pitchblack night. 'There was light' only after the landing of the cosmic vehicle on the planet and then the unknown beings experienced day and night, and life could begin and intelligence originate at the goal of their journey—in answer to a word of command.

In nearly all known creation legends the primordial truth is repeated that the word came from the light. There was a rich oral tradition on the Polynesian islands long before the first white man landed. A select circle of priests watched carefully to see that not a word of the old philosophical and astronomical wisdom was changed, but western civilisation and Christian missionaries stifled the rich tradition that the original population had possessed. In 1930 the Bishop Museum of Honolulu, which was the largest Polynesian collection in the world, sent two expeditions to the islands. Their aim was to safeguard the genealogies and songs that had survived the dubious blessing of western colonisation. Years later the Swedish scholar Bengt Danielsson, who had crossed the Pacific on the Kon Tiki raft with Thor Heyerdahl, visited some of the South Sea Islands with his wife and wrote down the traditions that were still alive in the consciousness of the islanders.

On the little island of Raroia in the Tuamotu group in the Pacific Ocean, 450 nautical miles north-east of Tahiti, Danielsson met an old sage whose name was Te-Yho-a-te-Pange. Danielsson tells us how this priest droned out the history of his people like a gramophone record. It is staggering:

'In the beginning there was only empty space, neither darkness nor light, neither land nor sea, neither sun nor sky. Everything was a big silent void. Untold ages went by...'

Could the account be more pertinent? Do we have to leave it to a primitive man in a loincloth, who lives on coconuts and fish, and has absolutely no technical knowledge, to explain to us what it looks like in space? But let Te-Yho-a-te-Pange go on talking:

'...Then the void began to move and turned into Po. Everything was still dark, very dark, then Po itself began to revolve...'

Have we reached the solar system now, have we entered the field of the orbits of the planets? (The void began to move.) Darkness still reigned. A sphere—called Po here—

became visible. It begins to revolve.

'New strange forces were at work. The night was transformed...'

A telling description. Now the attraction of the planet is at work (... new strange forces ...). We are sinking into the atmosphere. It grows bright as day.

'...the new matter was like sand, the sand became firm ground that grew upwards. Lastly "Papa", the earth mother, revealed herself and spread abroad and became a great country...'

Then people were on terra firma, which extended far and wide. But before they reached the earth's surface, which 'grew upwards' (an impression that arises when one comes on it from above), matter that was 'like sand' had to be traversed. Is that another way of describing the powerful frictional forces that the envelope of air exerts on the exterior of the space-ship?

Te-Yho-a-te-Pange continues:

'...there were plants, animals and fish in the water and they multiplied. The only thing that was lacking was man. Then Tangaloa created "Tiki", who was our first ancestor...'

We should never forget this myth of the creation. Perhaps it would be a good thing to tell it in our schools.

The *Popol Vuh* contained another wonderful account. This book, which is one of the 'great writings of the dawn of mankind' (Cordan) and is in the nature of a secret book was the holy scripture of the Quiché—Indians of the great Mayan family around Lake Atitlan in the Central American state of Guatemala.

Its comprehensive creation myth claims that men only partially stem from this earth, that 'gods' created the 'first beings endowed with reason', but destroyed all the unsuccessful examples and after performing their earthly tasks rose into heaven again, to the place where the 'heaven's heart' is, namely to Dabavil, to him 'who sees in the darkness'.

Is this the reason why the Quiché Indians were imbued with the concept of gods who dwelt in stone spheres and who could emerge from the stone? Does the ball game cult of this tribe, of which the *Popol Vuh* tells, have its roots in this creation myth? The ball game as cosmical and magical rite, as symbol of the flight to the stars?

Among the creation stories that strengthen my theory, another myth—that of the Chibcha (i.e. men)—is a real jewel. The historical home of these people, whom the Spaniards discovered in 1538, is on the east Colombian plateau.

The Spanish chronicler Pedro Simon describes the myth of the Chibcha in his *Noticias historiales de las conquistas de tierra firme en las Indias Occidentales*:

'It was night. There was still something of the world. The light was closed up in a big "something house" and came out of it. This "something house" is "Chimini-gagua", and it hid the light in it, so that it came out. In the brightness of the light things began to come into being ...'

I can see that it must have been difficult for translators and interpreters to find a clear-cut equivalent for the word 'something house'. But how lucky for us that they left this concept that is so hard to understand and did not replace it by a fanciful synonym. Otherwise we might not be able correctly to interpret the significance of this tradition and grasp its full importance. But now we can measure the 'something house' against our present knowledge. As the Chibchas had never seen a space-craft before, they obviously did not know what to call it. So they paraphrased it in words that were familiar to them: something like a house had landed and the 'gods' came out of it.

The traditions of the Incas in Peru say that even before the world was created a man named Uiracocha existed (i.e.

Viracocha, later the god Quetzalcoatl), whose full name Uiracocha Tachayachachic means 'Creator of the world things'. This god had originally been both man and woman. He settled in Tiahuanaco and created a race of giants there.

Is there perhaps a direct relation between the monolith in Tiahuanaco, the magnificent Gate of the Sun, and the traditional story of the creation? And are we interpreting the saga of the golden egg that came from the cosmos and whose passengers began the creation of men too arbitrarily if we take it at its face value, as an authentic account of a space-craft from unknown stars?

This golden or gleaming egg that fell from heaven is a veritable leitmotiv in mankind's traditional stories of creation.

A stylised drawing of a spherical space-ship on a ceremonial vessel (National Anthropological Museum, Mexico).

On Easter Island the gods were worshipped as 'lords of space'. Among them, Makemake is the god of the 'dwellers in the air'. His symbol is the egg!

There are two strange books in Tibet called Kantyua and Tantyua. Actually one cannot really speak of books in their case, for Kantyua alone comprises 108 parchment volumes which number 1,083 books in nine large divisions. Kantyua means 'the translated word of Buddha' and the sacred texts of Lamaism are collected in it. Kantyua has the same kind of importance as the Koran has for Islam. Tantyua means 'the translated doctrine' and is a 225-volume commentary on Kantyua. These Chinese printed books take up so much room that they are preserved in the cellars of several villages that lie hidden in the mountain valleys of Tibet. The separate parts of the texts are carved on wooden blocks 3 ft long, 4–8 ins thick and 6 ins wide. Since not more than eight blocks can go on one parchment page it is understandable that the volumes have to be housed in the cellars of whole villages. Only a hundredth part of these texts, whose original date is not fixed, has been translated. In both these mysterious works there is constant mention of 'pearls in the sky' and transparent spheres, in which the gods dwell, to show themselves to men at great intervals. If there was purposeful and coordinated research on Kantyua and Tantyua we should probably learn a very great deal about the 'gods' and their former activities on earth.

In the Indian world the Rigveda is considered to be the oldest book. The Song of Creation that it tells returns once more to that state of weightlessness and soundlessness that reigns in the infinity of the universe. I quote from Paul Frischauer's book *It is Written*:

'In those days there was neither not-being nor being. Neither the atmosphere nor the sky was above. What flew to and from Where? In whose keeping? What was the unfathomable? ... In those times there was neither death nor immortality. There was not a sign of day and night. This *one* breathed according to its own law without currents of air. Everything else but this was not present. In the beginning darkness was hidden in darkness ... The life-powerful that was enclosed by the void, the *one*, was born by the might of its hot urgency ...

'Was there an above, was there a below? ... Who knows for sure, who can say here whence they originated, whence this creation came?'

We should take especial notice of the phrase 'the life-powerful . . . was enclosed by the void'. As twentieth-century men we can hardly recognise this Song of Creation as anything but an account of a space journey.

But who can explain convincingly why ancient peoples all over the world, who did not know of each other's existence, told stories of the creation with the same basic core?

In the old Chinese classic Tao-te-king there is one of the most beautiful definitions of the origin of the cosmos, life and earth:

'The meaning that one can invent is not the eternal meaning. The name that one can name is not the eternal name. Beyond the nameable lies the beginning of the world. On this side of the nameable lies the birth of creatures.'

According to this definition, too, the 'beginning of the world' lies outside our spheres; on this side, 'this side of the nameable', lies only the 'birth or creatures'.

Egyptian priests provided the mummified dead in the tomb with texts containing instructions for their future behaviour on the other side. These books of the dead were very detailed; they contained advice covering every conceivable situation. The directives were meant to lead to reunion with the god Ptah. One of the oldest prayers in an Egyptian Book of the Dead runs:

> *'O world-egg, hear me.*
> *I am Horus of millions of years.*
> *I am lord and master of the throne.*
> *Freed from evil, I traverse the ages*
> *and spaces that are endless.'*

I am always delighted when I can 'prove' textual interpretations with pictures or, better still, tangible works by stonemasons. And circles, spheres and balls can be found in abundance. In the Tassili mountains in the Algerian Sahara figures in strange suits can be seen painted in hundreds of places on the rock face. They wear round helmets with antennae on their heads and seem to be floating weightlessly in space. I should make special mention of the Tassili sphere, which the Frenchman Henri Lhote discovered on the underside of a semi-circular rock. In a group of floating couples—a woman is pulling a man behind her—a sphere with four concentric circles is clearly visible. On the upper edge of the sphere a hatch is open and from it a thoroughly modern-looking TV aerial protrudes. From the right half of the sphere stretch two unmistakable hands with outspread fingers. Five floating figures who accompany the sphere wear tight-fitting helmets on their heads, white with red dots and red with white dots. Are they astronauts' helmets?

Another figure from Val Camonica in Italy. The headgear looks very much like some kind of aerial. A cosmonaut?

If today we were to give children a box of crayons and ask them to draw the moon flight out of their heads, the result would probably be something quite similar to the Tassili paintings. For the 'savages' who painted these memories of the visit of the 'gods' on the rock faces were probably at the mental age of a child.

81

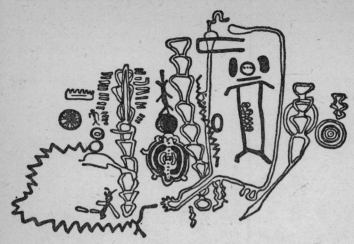

So far no one has given a plausible explanation of these complicated rock paintings from Santa Barbara in California. Note the different globular figures.

According to archaeologists these rock drawings from Invo County, California are gods. Is this really convincing? They seem more technical than divine.

The Tassili sphere is not the only piece of proof I have come across. Anyone who arrives in one of the districts in the following list should have a film in his camera, for he will be able to photograph spheres and circles galore, and have a good think about their origin. The list is, of course, only a brief selection.

Detail from an Assyrian cylinder seal. Behind the figure there appears to be a representation of a planetary system.

Kivik, Sweden, some 50 miles south of Simrishamn. In the celebrated rock tomb, which has a star in every guidebook, there are large numbers of plain circles and circles divided by a vertical line—all symbols of the gods.

Tanum, Sweden, north of Gothenburg. Several wonderful spheres and circles surrounded by rays.

Val Camonica, Italy, near Brescia. About 20,000 prehistoric paintings, including numerous radiating circles and 'gods' with helmets.

Fuencaliente, Spain, about 40 miles north-east of Cordoba. Many circles and spheres with and without a crown of rays.

Santa Barbara, USA, 50 miles north-east of Los Angeles. Partly interlaced circles with rays.

Inyo County, USA, east California on China Lake. Rings, stars, spheres, many-coloured rays and figures of 'gods'.

Circles and spheres, apparently strategically distributed, are found in countless places throughout the world.

To sum up: all spheres and circles—whether in creation myths, prehistoric drawings or later reliefs and paintings—

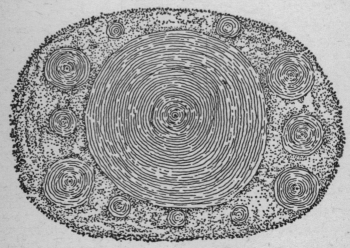

A ceremonial wood plaque from central Australia now in the National Museum of Victoria. Is it a stylised prehistoric picture of the world or a drawing of a planetary system?

represent 'god' or the 'godhead'. The rays are generally directed earthwards. In my opinion, this universal custom should give us something to think about.

I am convinced that the traditional spheres and divine eggs have more than a mere religious and symbolical significance. The time has come to look at these signs from another point of view. The patterns of thought we have followed so far may be absolutely wrong. So far we have lacked the pre-requisites fully to grasp the legacy of the 'gods' contained in the monuments and documents of our primitive ancestors. But today, when man has already set foot on the moon, he should no longer be satisfied with explanations that were coined in centuries when the theory of life was still firmly established and man was convinced he was the lord of creation!

Ironically enough, I may mention that prehistoric finds were excavated over a distance of 1,200 ft at Carschenna above Thusis in the parish of Sils in the canton of Graubunden, barely twenty miles from my home town. What has come to light so far? Rock faces and plaques covered with large numbers of spheres, circles, spirals and circles with rays. Why did I have to travel round the world, when the proofs of my theory were right outside my front door?

Spheres surrounded by rays, eggs and flying spheres are not only found on cave walls and cliffs, ancient stone reliefs and cylinder seals, but also in the round, made of hard stone, in many different parts of the world—generally scattered indiscriminately and in inhospitable country. In the USA, for example, balls have been found in Tennessee, Arizona, California and Ohio.

In 1940, Professor Marcel Homet, the archaeologist now living in Stuttgart and author of the well-known book *Sons of the Sun*, discovered a gigantic stone egg 328 ft long and 98 ft high on the upper Rio Branco in North Amazonas,

Brazil. On this colossal block, which was called Piedra Pintada (painted stone), Homet found countless letters, crosses and sun symbols over a surface area of some 700 sq yards. He assured me in conversation that there was not the slightest doubt that this magnificent specimen was no freak of nature, but stonemason's work carried out by countless hands over many decades.

Detail of the 'flying machine' shown on the Assyrian cylinder seal (see illustration facing page 144).

But the real archaeological sensation in the ball line awaits its solution in the small Central American state of Costa Rica. There hundreds, if not thousands, of artificial stone balls lie about in the middle of the jungle and on high mountains, in river deltas and on hilltops. Their diameters vary between a few inches and eight feet. The heaviest ball excavated to date weighs eight tons!

I had heard about this sensation and because of it I spent ten days in Costa Rica, a typical developing country, that has so far been shunned by the vast mass of tourists. My journey turned out to be anything but a pleasure trip, but all the hardships were richly rewarded by what I saw.

The first balls I came across were lying around in flat country for no apparent reason. Then I found several groups of balls on the tops of hills. Some always lay in the centre of the hill's longitudinal axis. I waded through the mud of a riverbed and found great groups of balls in strange formations that were unintelligible, although they must have been deliberate.

Forty-five balls have been lying in the burning sun of the white-hot Diquis plain since time immemorial. Have they something to say that we were or are incapable of understanding?

In order to satisfy my curiosity to see and photograph the balls near Piedras Blancas south-east of the River Coto, also in Costa Rica, we had to spend a whole day in a Land Rover to cover a distance of only sixty miles. Time and again we had to remove obstacles from the track, pull the Land Rover out of ruts and grind round innumerable bends. Finally our vehicle would take us no further. Bubu, a half-breed, who was guiding us, ran ahead of us for an hour and cleared the way of creatures. Without his precautions we should twice have run into spiders' webs of a size you simply cannot imagine. The poisonous bite of these loathsome creatures can be fatal.

At last we stood before two enormous balls, both taller than we were, in the midst of the virgin forest. It was precisely because the stones near Piedras Blancas lay deep in the jungle that I had wanted to see them with my own eyes. It is said that these balls are only a few hundred years old. No one who has stood there as I have could believe that. The jungle itself is primaeval and I am convinced that the balls must have lain there before the luxuriant vegetation began to thrive.

Today we have managed to 'transplant' Abu Simbel to another site using all kinds of modern machinery, but I doubt whether we could deposit balls like these in a primaeval forest.

I saw still more balls in Costa Rica.

Fifteen giant balls lie in a dead straight line in Golfo Dulce.

I found twelve balls near the village of Uvita north of the Sierra Brunquera.

Four balls have been excavated from the muddy bed of the River Esquina.

There are two balls on Camaronal Island and several balls on the summit of the Cordillera Brunquera in the neighbourhood of the River Diquis.

Most of these mysterious balls are made of granite or lava. There is little chance nowadays of finding out the exact number of stone balls that once existed. Today many fine specimens decorate gardens and parks and public buildings. Since one ancient saga also related that gold could be found in the middle of the balls, many of them have been hacked to bits with hammer and chisel. A strange thing is that there are no quarries for producing the balls anywhere near the sites where they have been found. As in other places any trace that could lead us back to the 'manufacturers' is missing.

During the clearing of woods and swamps at the foot of the Cordillera Brunquera in the Rio Diquis district in 1940 and 1941, the archaeologist Doris Z. Stone discovered several artificial stones. She wrote a detailed account of them that closes with the resigned remark: 'The balls of Costa Rica must be numbered among the unsolved megalithic puzzles of the world.'

In fact, we do not know who made the stone balls; we do not know what tools were used for the work; we do not know for what purpose the balls were cut out of the granite and we do not know when they were made. Everything that archaeologists say today in explanation of the Indian balls or sky balls, as the natives call them, is pure speculation. A local legend says that each ball represents the sun—an acceptable interpretation. But the archaeologists reject this version, because in these latitudes the sun has been represented in all ages as a golden orb, wheel or disc, and never as a ball—neither among the Incas, Mayas nor Aztecs.

One thing is quite certain. The stone balls cannot have originated without mechanical help. They are perfectly executed—absolutely spherical, with smoothly polished surfaces.

Archaeologists who have investigated the balls of Costa Rica confirm that none of them deviates in the slightest from a given diameter. This precision implies that the men who made them had a good knowledge of geometry and possessed the appropriate technical implements.

If the stonemasons had first buried the raw material in the earth and then worked on the protruding section, unevenness and inaccuracies would inevitably have resulted because the distances to the part stuck in the ground could no longer have been checked. This primitive procedure is out of the question. The raw material must have been transported from somewhere, because there are no nearby quarries, and that alone must have been very arduous. In addition, the stone blocks must have been broken out or cut out of the rock. My conclusion is that many workers were engaged on the task for a long time and that they possessed tools which made possible such perfect stone dressing.

Even if all this is accepted, it still does not explain why the finished balls were rolled to a particular site, e.g. the top of a mountain. What an absurd idea and what a tremendous expenditure of labour! However, an explanation is given, though it only seems suitable for the most superficial kind of guide-book. The gigantic balls were rolled down riverbeds. I should laugh at such naïvety if the problem involved were not so serious to me. The massive heavy balls would simply have stuck in the muddy, and in parts gravelly, riverbeds.

One irritating fact which cannot have altered in the course of the ages confronts the holders of the riverbed theory. Between the granite mountains in which the material for the majority of the balls must have been quarried and the sites where the balls were found in the Diquis delta the steaming jungle extends far and wide, and the three small rivers that exist are considerable obstacles to transporting material on such a scale without deeploading lorries, cranes and special freight ships. And as if these barriers were not enough, when seen from the granite cliffs most of the balls lie on the far side of the Rio Diquis! In other words, the forwarding agents would have had to 'conjure' the material over this barrier, too. I have noticed that whenever archaeologists cannot explain gigantic feats of transport, they have recourse to the so-called 'rolling theory'. But this is pitifully inadequate when one sees the giant balls on the tops of mountains. An expert has told me that at least twenty-four tons of raw material are needed to make a stone ball weighing sixteen tons. In view of the large numbers of balls, one can guess roughly what masses of raw material must have been moved about here in the past.

I had seen the miraculous world of the stone balls and convinced myself of their disturbing existence. Now I wanted to try to find the solution of this puzzle as well, but when I asked the Costa Ricans about the origin and meaning of the stone balls, I met silence and suspicion. Although visited by missionaries and 'enlightened' by continuous economic contacts with the west, the natives remained superstitious in their heart of hearts. Two archaeologists whom I questioned in the Museo Nacional of San José explained that the creation of the balls was connected with a star cult, perhaps too with calendar representations, and possibly with religious or magic signs. I needled away patiently, because these explanations did not satisfy me, but finally had to realise that the mystery of the balls was taboo to them, for some reason incomprehensible to me.

As the archaeologists could not or would not help any further, I tried asking some of the Indians. Trained by my acquaintance with natives in many countries, I soon sensed that they were afraid of something as soon as the conversation came round to the balls.

Nevertheless, it is extremely surprising that these poor creatures, who haggle over every centimo, would not guide me to a 1,800 foot-high cliff with three balls on top, no matter how much I offered them. Bubu was an exception.

A German, who has owned the Pension Anna in San José for over forty years, is considered to be the man who possesses most material about the balls. He pulled out many impressive pictures, but behaved as if he had to keep the secret of some buried treasure. He showed me sketches of formations and groupings of balls, but refused to give their exact location. I was not even allowed to copy his sketches. 'No, it's out of the question,' was his inevitable reply.

Even if I had not known it beforehand, I should have realised during my stay in Costa Rica that there is a mystery about the stone balls. I could not solve it, but my suspicion increased that the prehistoric balls and all the pictures of them in reliefs and on cave walls are directly linked with the visit of unknown intelligences, of intelligences who landed on our planet in a ball. They already knew and had proved that the sphere is the most suitable shape for interstellar space flights.

One day in the not too distant future the long journey back to the stars will start from our planet, and probably in spherical space-craft, because the sphere is the most natural of all shapes for the flight into the universe.

## 6

## THE SCIENCE-FICTION OF YESTERDAY
## IS TOMORROW'S REALITY

In *Chariots of the Gods?* I wrote a chapter in which I predicted a mass exodus of people from our planet to another heavenly body. I thought that this fantastic idea of mine was one way of easing the murderous population explosion, from which there seems to be no escape. In the end I cut this vision of the future out of my manuscript before it went to the printers. I did not want to frighten my readers with such 'impossible' ideas. But progress has caught up with me; I should have had more confidence and left it in.

In the interim there have been Russian and American experiments whose ultimate aim is to put this idea into practice, even though it sounds like a science-fiction project today. Professor Carl Sagan of Harvard and Professor Dmitri Martinov of the Sternberg Institute in Moscow are both doing research work along the same lines. They want to conquer Venus for mankind—Venus, whose distance from the earth varies between 26,000,000 miles (inferior conjunction) and 160,625,000 miles (superior conjunction).

For research in the laboratory they have at their disposal 'reconnaissance reports' from the Russian Venus sondes and the American *Mariner*. On 6 June, 1969, TASS gave the surface temperature of Venus as varying between 400° and 530° C. This agrees approximately with the 1967 reports of the American *Mariner V*, which radioed back to its ground station a temperature of 480° C and an atmospheric pressure of 50 to 70. The Russians also received details from the sondes which had made soft landings. According to them, the Venusian atmosphere has a carbon dioxide content of 93 to 97%, while nitrogen makes up 2 to 5% and oxygen apparently only accounts for 0·4%. At a pressure of barely 1 atmosphere the measuring apparatus registered a water content of only 4 to 11 milligrams per litre.

These data are valuable working material. On the basis of them both Martinov and Sagan made plans for opening up the morning and evening star biologically. Carl Sagan published his ideas in the scientific periodical *Science*, which has the reputation of only publishing contributions that have been thoroughly examined and have stood up to strict scientific scrutiny.

Sagan thinks that in the near future—he is speaking of a few decades—space-ships with big cargo holds will unload thousands of tons of blue algae into the Venusian atmosphere, i.e. 'blow' them towards the surface of Venus. Blue algae stay alive even at high temperatures, but reduce the high proportion of carbon dioxide by their metabolism. Owing to this steady reduction of carbon dioxide the surface temperature would gradually fall and finally sink below 100° C. Blue algae would then cause the same chemical reaction as once took place in the 'primitive soup' of our earth. With the help of light and water carbon dioxide particles could be transformed into oxygen. But once the blue algae had lowered the temperature below 100° C, a rain like the Flood would fall on Venus. Light, oxygen and water would then provide the prerequisites for the first primitive life.

Since scientists have already thought of evacuating mankind to another planet, they have also planned protective measures for we sensitive delicate creatures. In the second phase of their colonisation of Venus chemicals would be sprayed to destroy micro-organisms that might be dangerous to the lord of creation.

Only very distant generations will live to see the execution of this gigantic project. For although such plans can be speeded up by modern technological development, we must allow a considerable period of time for the evolution of the new world. At present scientists talk of 1,000 years before the first evacuation space-ship can travel to Venus.

Technology pampers us. On 20 July, 1969, at 0300 hours, 56 minutes 20 seconds Central European Time, hundreds of millions of people saw the two astronauts Neil Alden Armstrong and Edwin E. Aldrin become the first men to set foot on the moon. This was the most magnificent achievement in space travel to date and it fascinated and astonished humanity all over the globe. But even while man was following the breathtaking flight to the moon, science was already occupied with exploratory flights to Mars and Venus, and even with a vast human migration to earth's sister planet. Just as the conquest of the moon began with unmanned satellites, Venus is now being investigated with unmanned sondes. On 18 May, 1969, Moscow reported that after a flight lasting 130 days the Venus sonde 5 had ended the journey of 156,250,000 miles with a payload of 2,260 lb. When the sonde was still 31,250 miles from Venus, the ground station radioed the last command. Then the sonde sent down a capsule full of instruments by parachute. TASS stated that the parachute drop lasted 53 minutes.

The distance of Venus from the earth depends on the distance of its orbit from the earth, and it varies between 26,000,000 and 160,625,000 miles. The Russian sondes do not reach Venus by the shortest route. That sounds paradoxical, but the Russian principle for the flight paths of the Venus sondes is valid for all present-day interplanetary flights. The flight path is calculated according to the minimum quantity of fuel necessary for the transport of the space-craft. If the sonde were launched on the direct course to Venus, it would have to have an initial velocity of 20 miles per second. In that case vast quantities of fuel would be used not only at launching, but also for braking the initial velocity later on. Consequently ballistic experts prefer to calculate flight paths that approximate to the movement of the earth as far as possible. The most favourable route on these premises is ten times as long as the direct route, but it allows a launching velocity of 7.175 miles per second and a far lower consumption of fuel.

What is there left that is really Utopian? Preliminary research becomes applied science at such a breathtaking speed that science-fiction writers have a hard job inventing the unimaginable.

In 1969, Professor Hans Laven, Director of the Institute of Genetics at the University of Mainz, published a report according to which millions of insects that are dangerous to men, animals and plants as disease carriers could be killed without the use of insecticides, the chemical agents with which dangerous insects and their brood are destroyed at present. As early as 1967 Laven was able to show the efficacy of his research in the Burmese town of Okpo, which was plagued by midges. Within a few months Okpo was freed of its plague.

Laven had experimented in the laboratories at Mainz for years. In the process he found out that there is a natural incompatibility between midges of different origin. Midges from North Germany showed themselves willing to mate with members of their species from Schwaben, but the midge offspring produced were not viable. If the midges from different German provinces had an antipathy for each other, then midges from different continents would surely produce offspring that were unviable. So they bred a race from California and French midges. When they were released in the town of Okpo, the males of the Mainz bastard race proved to be great lovers and competed successfully with the Burmese males, but the eggs laid by the wives they had mated with did not produce any new midges. The chromosome count of the different species of midge did not tally and genetic destruction took place. The advantage of this way of destroying midges is obvious—the risk of contaminating food and plants by spraying them with insecticides disappears.

Professor Laven is continuing his researches on the basis of the most recent genetic discoveries. He radiates male midges with about 4,000 r X-rays. This does not cause the creatures any organic damage, but the chromosome chain between the genes is damaged. The chromosome household is disturbed, the genes are changed and insects develop in an unprogrammed sequence. They are still capable of reproduction, but their offspring are reduced in every way i.e. in number, size, etc. Of some midge generations treated in this way who continue to hand down their planned disability, Laven said: 'There is no cure for semi-sterility, because it is hereditary.'

Laven is convinced that in a comparatively short time it will be possible to use his model experiment against other harmful insects; he even thinks that the plagues of rats in many parts of the world can be tackled in this way.

The tremendous possibilities of manipulating the genetic code are no visionary dream. We are dealing with scientific facts. Of course between yesterday and tomorrow lies the 'abyss' that will be crossed. Most probably we shall only rediscover something that has already happened.

One day new knowledge and techniques will create a human organism suited for interstellar flight, one that will not fall sick and will be equal to all the burdens and stresses imposed on it.

Medical science has been performing transplants of

RIGHT: This massive block of stone has sharp-edged grooves that could not have been made with stone axes or wooden wedges.

BOTTOM LEFT: This statue from Tiahuanaco, carved out of a single block, stands in La Paz (Bolivia) today. Who made such huge monuments? Are they likenesses of extraterrestrial beings?

BOTTOM RIGHT: Fragment of a Tiahuanaco statue of a quite different kind. Today it, too, is in La Paz (Bolivia).

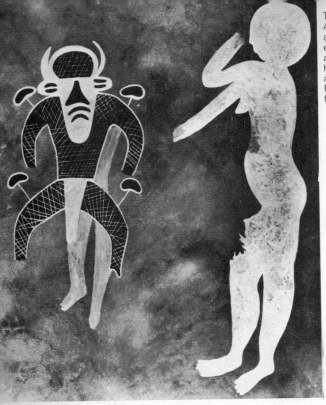

This man from Auanrhet, Tassili, antenna-like excrescences on and thighs. His he has slits for eyes, and mouth.
RIGHT: A naked female figure.

Cajamarquilla near Lima (Peru). Fox-holes? Grain silos? There are 209 of them in a straight line. Why holes which everyone was bound to fall into?

Close-up of a hole. Diameter 23 ins, dep 5 ft 7 ins.

ABOVE: Aztec ceremonial disc of serpentine. Sun
god or theologically exaggerated picture of a
cosmonaut?
RIGHT: Milestone of King Melichkhon with sun,
moon and a beautifully carved round body. Earth?
Venus? Or a space sphere?

BELOW: Rock painting from Auanrhet, Tassili, about 8,000 years old, with strange
figures. The open hatch and the two protrusions on the right of the spherical object raise
problems.

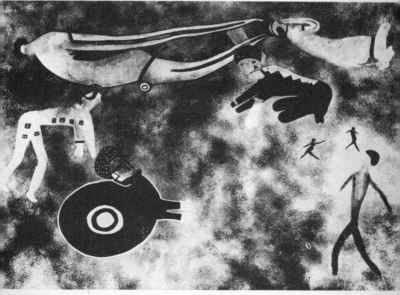

Henri Lhote called this nearly 19-foot high figure from Yabbaren, Tassili, the great martian god. He looks every inch a cosmonaut even compared with our own moon travellers!

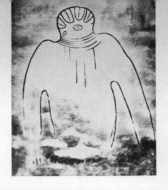

ABOVE: This rock painting was found 25 miles south of Fergana (Uzbek, USSR).

ABOVE RIGHT: Space travellers from a rock drawing in Val Camonica (Italy).

Circular Mayan calendar. Where did the Mayas get their astronomic and mathematical knowledge from? Has the shape any relation to its content?

organs for more than twenty years, but unproductive and unpleasant sensation-mongering about these important scientific operations did not begin until after the first heart transplant. When pieces of skin were transplanted in the 1940's, when bones were changed in 1948 and when a kidney was transplanted in 1950, there was not a murmur. In 1954 the first transplant of a limb on to a dog was successful. In 1955 somebody else's lung was inserted into a man. In 1967 a pancreas was transplanted and in 1969 doctors risked the transplant of a liver. There were successful transplants of other organs besides these.

It was the heart transplant which first unleashed lively discussions and violent opposition in all the newspapers in the world—probably because there is an unspoken feeling that it fulfills more than the function of a simple pump. Strangely enough, man, who enjoys living and is terrified of death, has not welcomed this advance of medical science with open arms. Yet the prospect of being able to save a man's life by changing a defective organ is an important one. Many teams of surgeons know how to perform the surgical operation. As soon as the threshold of the immune reaction can be lowered without the body's own defences against infection being endangered, transplants should take place as a matter of course, just like operations for appendicitis. But that is the moment when the supply of spare organs will cause difficulties. So as to be independent of family and religious taboos when vital operations have to be performed, we shall collect human organs for future recipients in 'organ banks'. Today there are blood banks in hospitals all over the world. Why has no one made a fuss about them? For blood is still the sap of life and far more mysterious than the heart pump. Of course, blood is donated voluntarily. Why should not this ultimately be true of organs, given by men who know that they are going to die or by their relatives?

I believe that organ transplants, too, are only a transitional stage. If scientists eventually succeed in programming the DNA double helix in the nucleus with information for the construction or reconstruction of organs, the Frankenstein methods will soon be forgotten. The Russian scientist L. V. Poleschayev is already able to make a damaged scalp regenerate itself independently and he has even managed to make amputated limbs start to grow again. One day there will be gene surgery as well. Sheer fantasy? I do not think so, because I know that Doctor Teh Ping Lin of San Francisco succeeded in giving a mouse's egg an injection as early as 1966. A mouse's egg is only one tenth the size of a red corpuscle and cannot be seen at all with the naked eye!

Professor E. H. Graul, Director of the Institute for Radiobiology and the Medical Use of Isotopes at Philips University in Marburg, and the cyberneticist Dr Herbert W. Franke gave a forecast of medicine and its fringe areas in the years 1985 and 2000 in the *Deutschen Ärzteblatt*:

Forecast for the year 1985.

Mastery of transplating animal and human organs, elimination of immune reactions.

Routine use of artificially made organs and/or biological systems (artificial organs made of plastics and/or electronic components—on cyborg lines).

Marked progress in the gerontological and geriatric field. The average expectation of life is about 85 years.

The ageing process will be manipulated advantageously, the degeneration inherent in old age will be slowed down, both physically and psychically.

The first positive conclusions about the production of primitive forms of artificial life.

Biomedical electronics will influence practical medicine (for example, electronic artificial limbs, radar for the blind, limbs with servomechanisms and many other things).

Forecast for the year 2000.

Deep-freezing of men for hours or days.

Determination of a child's sex before birth.

Possibilities of transplanting all organs.

The correction of hereditary defects.

Continuous genetic manipulation of animals and plants.

Production of artificial forms of primitive life.

Use of X-ray and gamma type Laser beams.

The centre of this rock drawing from Toro Muerto (Peru) looks like an X-ray plate of the thorax. Its meaning is unknown.

General biochemical immunisation against diseases.

General diffusion of the cyborg technique (artificial organs).

Manipulation of organisms by electric stimulation of the brain.

Drugs for the control of man's emotional states, chemical aids for the improvement of memory and the ability to learn.

I suggest that unknown intelligences were able to do all these things in the dim mists of time.

I suggest that the 'gods' left this knowledge behind when they visited earth.

I suggest that discoveries, which still lie ahead of us in

the broad field of research, have been stored in the human memory since time immemorial and are only waiting to be summoned up again.

The experiments of David E. Bressler of the University of Los Angeles and Morton Edward Bittermans of Bryn Mawr College in Pennsylvania are a step along this road. They implanted additional brain cells into fish. The fish enriched with the transplanted brain substance soon proved to be much more intelligent than their untreated fellows. Cleveland Hospital (USA) is conducting a series of experiments in which monkey's brains are put into dogs.

Why did the Mayan priests tear the beating hearts out of their prisoners' breasts?

Why were cannibals convinced that they took over the strength and intelligence of their dead enemies when they ate them?

Why does a myth from remote antiquity claim the body only belongs to man temporarily, and that he must give it back to his 'master' at any time?

Ought we to suspect that the human sacrifices practised for millennia were more than esoteric religious observances? Were they distorted memories of transplants, operations, cell regenerations, that were handed down for thousands of years in a terrifyingly garbled form?

Let us examine another possibility. The 'thinking' computer will also be useful to man in the peaceful conquest of the universe. However startling its calculating achievements may seem today, the supplying of information by the electronic wonder is still in its infancy.

About 200 years ago the brilliant mathematician Leonhard Euler calculated the constant pi for the area of a circle to 600 decimal places. This fabulous achievement took him several years. One of the first computers spat out the constant pi worked out to 2,000 decimal places in a few seconds. As a routine task a modern computer gives the constant pi to 100,000 decimal places in a nano second = 1 milliardth of a second.

Today the 'brain' of a computer, its central core, operates with about one million information units. In computer

Why did Mayan priests tear the hearts out of the living bodies of their victims? Was it perhaps the memory of a misunderstood technique of operating used by the 'gods'?

jargon they are called 'bits'. The human brain works in a very similar fashion. Molecular memory units and nerve switches store and process information. The child in the cradle already stores it—even if unconsciously. Throughout our life we store information, so that we can call on it when needed. Only too often we have to admit that our brain does not operate very reliably with the 'hoarded' knowledge.

The central core of a computer works with a precision of a quite different order. Our brain works with more than 15 milliard switches, whereas a large modern computer uses only 10 million. Further elements can be set up between the switches by cross connections. Then why does a computer work so much more reliably than our brain? Because generally nine-tenths of our brain lies fallow, but the computer always has all its bits present.

Even today the superiority of the computer is shame-making. If our brain is to work at its maximum, we have to concentrate on a specific problem. But the computer can solve millions of different problems at the same time.

The fastest calculator in use in Europe today is in the Institute for Plasma Physics at Garching, near Munich. It carries out 16·6 million calculating operations per second. In the machine's electronic stomach 750,000 transistors are connected with each other in the shortest way by means of photolithographically produced plug board diagrams. The electromagnetic waves that make the connections move with the speed of light. Computer people talk of access times of $1\frac{1}{2}$ nano seconds as mere routine. In $1\frac{1}{2}$ nano seconds a light ray covers 1 ft 6 ins.

But when we learn that the newest computer of the Data Control Corporation performs 36 million calculating operations a second, then the fastest machine in Europe is already a rather antiquated model. In comparison, one of the General Electric models, the GE-235, can be called a home computer. It only solves 165,000 problems a second, but you do not have to buy it. Its services can be hired for four cents per second.

The ferrite storage unit of a modern computer houses 200,000 digits in a space of one square metre. Magnetic tape storage units swallow up 10 million bits of data. And computers of all kinds are first-class pupils. They check themselves and never make the same mistake twice.

Today computers still need interpreters who translate our language, figures and concepts into the various computer languages. Direct conversation with the uncanny creature is expected by 1980. In America, but even more so in England, which is far advanced in computer techniques, they are trying to dissolve human speech into symbolic groups that the computer understands. This is the direction in which all computer manufacturers are working. But for IBM, the biggest producer of computers, language is far too slow a means of communication between man and machine. They are looking for another medium of transmitting information.

I have said that computer technology is only at the beginning of the great possibilities that lie ahead of it. Future research has an eerie goal: the biotronic storage unit. Nucleic acids appear to possess magnetic properties. Should this assumption prove to be correct, they would be the smallest information carriers of all. If these investigations reached their goal, the processing unit of a computer, which still takes up a lot of room, would be reduced to the size of the human brain. Biotronic information units would only be as big as chain molecules. I think that this line of research will ultimately succeed, but I am afraid that biotronic calculators would be susceptible to infection by viruses and bacteria.

Interstellar space travel operates with distances of hundreds of million of miles. At the speeds that are to be expected, the computer will be more than a mere ready reckoner. Even if computer manufacturers today resist the claim that computers will be able to think and act independently one fine day, that day will come. Then computers will steer space-ships between the planets on their own.

Far be it from me to assert that our ancestors knew anything about computers, integrated or electronic measuring apparatus. But since I am convinced that extraterrestrial intelligences visited the earth, their space-craft must have operated with suitable instruments. And since we men are programmed by the 'gods', we shall soon be masters of the same technological miracles.

# 7

# CONVERSATIONS IN MOSCOW

On Saturday, 18 May, 1968, Alexander Kassanzev, the famous Soviet writer, put the three sculptures which had impressed me so profoundly carefully back in the glass case opposite the window in his Moscow flat. They were old Japanese statues, cast in bronze, and they seemed to be dressed in space suits. The largest of the statues was nearly 2 ft high and had a diameter of about 5 ins. Tight-fitting bands ran from the shoulders, crossed the chest and joined up again between the thighs at the height of the buttocks. A broad belt with rivets on it was round the hips. The whole suit down to the knees was fitted with pocket-like protrusions. The helmet was tightly bound to the trunk with pads and bands. Comical-looking hollows appeared to be openings for built-in breathing or listening apparatus. I noticed two more openings on the lower half of the head.

But the most fascinating thing about the figures was undoubtedly the large glasses with lenses set at an angle. I could not see any weapons, unless the short staff in the gloved left hand could be called a weapon. 'A mini-laser beam,' the author of a science-fiction novel might say.

Agog with curiosity, I asked Kassanzev where these figures came from and who had given them to him.

He chuckled into his beard: 'A Japanese colleague gave them to me before the war, in the spring of 1939. The figures were found during excavations on the island of

Hondo in Japan. They are dated to long before our era. The figures have striking, indeed unmistakable, space-traveller characteristics, but no one can say how or why Japanese artists dressed their figurines in such suits. But one thing seems to be clear. Neither "snow goggles" nor lenses of this kind were known in ancient Japan.'

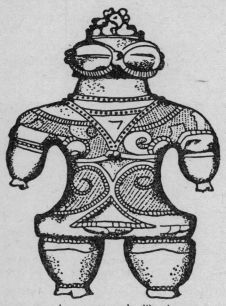

No one can say when snow-goggles like these were worn in Japan. Did the sculptor carve the image of a space traveller whom he had seen?

Then Alexander Kassanzev got into his decrepit old car and drove me through the splendid broad streets of Moscow to the Sternberg Institute of Moscow University, where he had arranged for me to meet Professor Josef Samuilovich Shklovsky, the Director of the Radio-astronomy Department.

What an experience it was, this Institute at Universitets-prospekt 13! It hummed like a beehive, teemed like an ant colony. The students' desks and tables stood higgledy-piggledy wherever there was room. Empty tin cans served

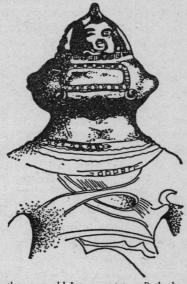

This is another very old Japanese statue. Both show markings which could be straps and fixing points for equipment.

as ashtrays. On the walls there were gigantic astronomical maps and in front of them students arguing. In one corner students struggled over a mathematical formula, in the opposite one others were busy with a complicated measuring apparatus. You felt instinctively that research was carried on here as teamwork.

The door to Professor Shklovsky's workroom was ajar, The room itself had the strange mixed smell of books, files and dust peculiar to rooms—as I have often noticed—in which the old is preserved and the new critically tested.

Professor Shklovsky rose from behind his massive desk, which was covered with printed papers and manuscripts, and greeted me with a suspicious laugh: 'So you're the Swiss!'

It sounded like a rebuke, as if the gaunt old man really meant: 'So you're the man who upsets the subjects of that quite peaceful country with your shocking theories.' As a result, our conversation, carried on in English, began with a certain amount of reserve. Calmly, deliberately, often looking for the right word, the famous man—and he knows that he is famous—explained his Martian moon theory. He thinks that the two moons of our neighbouring planet are artificial satellites. While he explained his arguments in favour of this theory, he would repeat in all modesty that it was only his private opinion.

After lunch in the crowded canteen, Professor Shklovsky relaxed his suspicious attitude a little. We began a lively argument about the impossible possibilities in the cosmos. In the end I had the satisfaction of realising that this leading expert of the eastern world, too, does not exclude the possibility of a visit in the past by unknown intelligences from the cosmos. He suspects that there are planets with intelligent life within a radius of 100 light years.

'But what about the distances, Professor? How will the incalculable distances between the stars be bridged?'

Shklovsky answered unhesitatingly: 'Obviously there is no cut and dried answer to that. Automatons or cybernetically controlled stations are not subject, as you know, to the calendar of 'normal' years. So what is there to stop a robot making a thousand year journey undamaged? After all, some of the satellites sent on their travels by us will still be functioning when we have long been in our graves.'

That is the opinion of a scientist who is an expert on the subject. It points to the technical possibility of bridging unimaginable distances. However, it still does not explain how and by what means intelligences could survive such periods of time.

My helpful friend Alexander Kassanzev was waiting for me in his old car. He had been talking to the students. The Institute is like home to him. Now he wanted to take me to the Pushkin Museum, which houses magnificent collections of Assyrian, Persian, Greek and Roman art. En route we spoke about a fascinating piece of Chinese research which would have made a great impression on our archaeologists. As we drove along Prouzenskaya Quay, Kassanzev told me many details of the latest conclusions Chinese scholars had reached and I spoke the key-words into my portable tape-recorder. When we stopped at traffic-lights as we did at Zoubovski Boulevard, I spelt out names and places clearly. The story I took home on tape was so exciting that it richly repaid the expense and efforts of the journey.

Kassanzev mostly talked about the remarkable finds in the Chinese mountains of Baian Kara Ula. His story sounded like a fairytale.

This Kassanzev's story: 'It was in 1938 that the Chinese archaeologist Chi Pu Tei discovered graves arranged in rows in the mountain caves of Baian Kara Ula in the Sino-Tibetan border district. He found small skeletons of beings with delicate frames, who nevertheless had rather large skulls. On the walls of the caves he discovered rock drawings which portrayed beings with round helmets. The stars, sun and moon were also scratched on the rock and connected by groups of pea-sized dots. Chi Pu Tei and his assistants managed—and this is the sensational thing about the find—to salvage 716 granite plates, which were 2 cm thick and looked very like our long-playing records. These stone plates had a hole in the middle from which a double-grooved incised script ran out spirally to the edge of the plate.

Chinese archaeologists knew that the Dropa and Kham

(Sikang) tribes had once lived in this deserted region. And anthropologists said of these mountain tribes that they had been of small stature, with an average height of only 5 ft 3 ins . . .'

'And how did they get the big skulls?'

'It was that part of the discovery that upset all previous anthropological classifications. Scholars could not place the high, broad skulls on the tiny skeletons of the Dropa and Kham. Not with the best will in the world. When Chi Pu Tei published his theory in 1940, he met with nothing but scorn. He claimed that the Dropas and Khams must have been an extinct species of mountain ape.'

'Then how did the stone plates originate? Are the apes supposed to have made them?'

'Of course not. In Chi Pu Tei's view they were placed in the caves by people from a later culture. On the face of it his theory did seem rather ridiculous. Who ever heard of rows of graves made by apes?'

'What happened next? Were the finds filed away in the archives of unexplained anthropologico-archaeological cases and forgotten?'

'Very nearly! For twenty years several clever men racked their brains to solve the riddle of the stone plates. Not until 1962 was Professor Tsum Um Nui of the Academy of Pre-historic Research in Peking able to decipher parts of the incised script.'

'What did they say?'

Kassanzev became serious: 'The story that was deci-phered was so hair-raising that at first the Academy for Prehistoric Research forbade Tsum Um Nui to publish his work at all.'

'And that was that?'

'Tsum Um Nui is a stubborn fellow; he went on work-ing doggedly. He could prove without any doubt that the incised script was not just a bad practical joke by some authority on prehistoric writing. For even serious scholars often show a sense of humour. Think of Piltdown Man. In cooperation with geologists he showed that the stone plates had a high cobalt and metal content. Physicists found out that all 716 plates had a high vibration rhythm, which led to the conclusion that they had been exposed to very high voltages at some time.'

Kassanzev turned off from Kropotkinskaya Quay and drove to the kerb of Volkhonka Street. The car stopped outside the Pushkin Museum. I was so entranced by his story that I wanted to hear the rest of it standing on the pavement, but Kassanzev took me by the arm and led me into the museum. We sat on a bench between tall glass cases.

'Please go on!'

'Tsum Um Nui now had four scientists who supported his theory. In 1963 he decided to publish it in spite of the Academy's doubts. I have heard that his publication was known to you in the west, but was not taken seriously. Over here, too, only a few courageous scholars took any notice of the stone plate theory. Very recently our philologist Dr Vyatcheslav Saizev published extracts from the stone plate story for the periodical *Sputnik*. The whole story is preserved in the Peking Academy and the historical archives of Taipeh in Formosa.'

'What is so explosive and shocking about the stone plate story?'

'The story is only upsetting and bizarre to people who are unwilling to face things that may throw new light on our origin. The stone plate story says that 12,000 years ago, reckoned from today, a group of their people had crashed on to the third planet in this system. Their aircraft—that is an exact translation of the groove hieroglyphs—had no longer had enough power to leave this world again. They had been destroyed in the remote and inaccessible mountains. There had been no means and materials for building new aircraft.'

'Is all that written on the stone plates?'

'Yes, and then we are told that these beings who had crashed on to the earth had tried to make friends with the inhabitants of the mountains, but had been hunted down and killed. The story ends almost literally: "Men, women and children hid themselves in the caves until sunset. Then they believed the sign and saw that the others had come with peaceful intentions this time . . ." That's more or less how it ends.'

'Is there anything else that corroborates the contents of the stone plate story?'

'There are the graves in rows, the rock drawings and the plates themselves. And there are also the Chinese sagas which precisely in the Baian Kara Ula region tell of small spindly yellow beings who came down from the clouds. The myth goes on to say that the alien creatures were shunned by the Dropas because of their ugliness, indeed that they were killed by the men in "the quick way".'

'Kassanzev, why isn't this fascinating story discussed all over the world? Is it really well enough known?'

My companion laughed, put his hand on my arm and said with mild resignation: 'Here in Moscow the story *is* known, you only need to keep your ears open. But the story contains too many facts which cannot be slotted straight into the painfully constructed chronology of archaeology and anthropology. Authorities who attach importance to their reputation in the scientific world would have to abandon a great deal of their own theories if they were to take serious notice of Baian Kara Ula. Surely it is very human to take the easy way out? To keep silent or laugh discreetly and condescendingly? When recognised scientists stick together and smile or say nothing, the boldest man loses enthusiasm in a subject that is too hot to handle.'

I am still too young to be able to resign myself. I believe in the disturbing power of ideas that cannot be hushed up.

# 8

## ANCIENT SITES THAT DESERVE INVESTIGATION

When I was in Peru in 1965, the closest view of the 820 foot-high three-armed trident on the cliffside in the Bay of Pisco was from a mile out at sea. On our journey in 1968, Hans Neuner and I planned to go ashore, free at least part of an arm from the sand and take photographs.

After an unsuccessful attempt to reach the three-armed trident by land in a rented car that kept on bogging down in the sand dunes, we persuaded a fisherman to take us into the bay. For a good two hours we swung along in a light breeze until the fisherman explained that he could go no nearer the coast because his boat would be in danger of being ripped apart by the sharp underwater reefs.

So we had no choice but to get into the water and wade and swim the remaining fifty yards to the shore with all our clothes on, including our shoes—because of the stinging fish. We pushed our tools, measuring tapes and cameras ahead of us, wrapped in plastic containers. When we reached the first coastal cliffs, we took off our wet things and tramped through the hot sand to the cliff face.

Unfortunately well-disposed gods do not lend super-natural powers to curious idealists. After a few hours of hard work we had to admit that it was beyond our powers to free even a single bit of the trident from the hard layer of sand.

Nevertheless, we took some accurate measurements that were worth our while. The individual arms of the trident are about 12 ft 6 ins wide. They consist of snow-white phosphorescent blocks that are as hard as granite. Before they were covered with sand, i.e. as long as the early inhabitants kept them clean, these dazzling, brilliant signals to the 'gods' must have 'shouted' to heaven.

There are some archaeologists who think that the trident on the cliffs of the Bay of Pisco was a landmark intended for shipping. The fact that the trident lies in a bay and cannot be seen from all sides by passing ships is against this theory. Another argument against it is that a landmark of this size would have been excessively large for coastal shipping, and the existence of deep sea shipping in remote antiquity is doubtful, to say the least. But the main thing against it is that the makers constructed their trident facing heavenwards. We might also ask why, if navigational marks were needed for some kind of shipping, the ancients did not make use of the two islands which lie far out at sea on the line of an extension of the central arm of the trident. They provided natural aids to orientation which would have been visible from afar to any ship regardless of the direction from which it approached the bay. So why a navigational mark that seafarers coming from either north or south could not see at all? And why one that points heavenwards? I may mention in passing—to clinch matters —that apart from a sandy desert there is absolutely nothing that could have attracted seafarers and that the waters with their sharp reefs must have been an unsuitable anchorage even in prehistoric times.

Another fact supports my theory about this signal that faces heavenwards. Only 100 miles away as the crow flies lies the plain of Nazca with its mysterious ground markings, which were only discovered in the 1930's. Since then archaeologists have been racking their brains over the

This sketch by Maria Reiche clarifies the relative size of the stylised figures on the pampa of Nazca. The biggest figure (not shown here) is about 27, yards long.

geometrical system of lines, animal drawings and neatly arranged bits of stone which extend over an area some 30 miles long between Palpa in the north and Nazca in the south. To me they look just like an airport lay-out.

Anyone who flies over the plain can see the lines shining up at him clearly—even from a great height they are perfectly recognisable. They stretch for miles, sometimes running parallel to each other, sometimes intersecting or joining up to form trapezoids with sides as long as 850 yards. Between these dead straight tracks it is possible to make out the outlines of enormous animal figures, the largest of which in its great extension measures some 275 yards.

Seen at close quarters the lines turn out to be deep furrows which expose the yellowish-white subsoil of the pampa and stand out clearly from the crustlike upper layer of brown desert sand. Maria Reiche, who has been working on the preservation, measurement and interpretation of the drawings since 1946 and was the first person to prepare field sketches with measuring tape and sextant of the triangles, rectangles, straight lines and numerous animal figures, found out later why the ground above the Ingenio valley was more suitable than any other for making clearly recognisable markings that would survive the centuries. The reason was that it only rains for an average of twenty minutes a year in the Nazca region. Otherwise a dry, hot climate prevails. Weathering is done by the sand-bearing wind which also carries away all loose material lying on the surface, leaving behind only pebbles which continually break up owing to the great differences in temperature. Moreover, the so-called 'desert varnish', which has a brown glint after oxidisation, has formed on them. In order to make the gigantic drawings show up, the constructors only had to remove the dark surface stones and scratch up the light subsoil of fine alluvial material.

But who made the pictures and why did they make them

so big that one can only get an overall impression of them from a great height, for example from an aircraft?

Did they already possess a highly developed system of surveying by means of which they transferred their small-scale plans to a gigantic scale with absolute accuracy?

Maria Reiche says: 'The designers, who could only have recognised the perfection of their own creations from the air, must have previously planned and drawn them on a smaller scale. How they were then able to put each line in its right place and alignment over large distances is a puzzle that will take us many years to solve.'

So far scholars have taken far too little notice of the phenomena on the pampa of Nazca. At first they thought that the dead straight lines were old Inca roads or irrigation channels. Both explanations are nonsensical! Why should 'roads' begin in the middle of a plain and then suddenly come to a stop? If the lines were roads, why should they intersect in a system of coordinates? And why are they laid out towards points of the compass when the purpose of roads is to reach earthly goals by the shortest possible route? And why should irrigation canals be in the form of birds, spiders and snakes?

Maria Reiche, who has worked the longest and most extensively on solving the secret of the Nazca plain and has written about it in her book *Secret of the Desert*, published in 1968, rejects these interpretations. She thinks it much more likely that the drawings are connected with the calendar, besides having a religious meaning. In her view the ground markings are observations of the heavens meant to be handed down to posterity in an indestructible form. However, she makes the following reservation: 'It is not absolutely certain if it is possible to interpret all the lines astronomically, for there are some (including many north–south lines) which could not have corresponded to any star appearing on the horizon during the period in question.

But if it was intended to draw the position of constellations not only on but also above the horizon, there would be so many possible explanations of the lines that it would be extremely difficult to prove any of them.'

I know that Maria Reiche does not share my interpretation of the geometrical drawings at Nazca, because the results of her investigations to date do not justify such daring conclusions. In spite of that I should like to be allowed to expound my theory.

At some time in the past unknown intelligences landed on the uninhabited plain near the present-day town of Nazca and built an improvised airfield for their spacecraft which were to operate in the vicinity of the earth. They laid down two runways on the ideal terrain. Or did they mark the landing strips with a raw material unknown to us? As on other occasions, the cosmonauts carried out their mission and returned to their own planet.

But the Pre-Inca tribes, who had observed these beings at work and been tremendously impressed by them, longed passionately for the return of these 'gods'. They waited for years and when their wish was not fulfilled, they began to make new lines on the plain, just as they had seen the 'gods' make them. That is how the extensions of the first two landing strips originated.

Yet still the 'gods' did not appear. What had gone wrong? How had they annoyed the 'heavenly ones'? A priest remembered that the 'gods' had come from the stars and advised them to lay the enticing lines out according to the stars. The work began again and the tracks laid out according to the constellations came into being.

But the 'gods' still stayed away.

In the interim, generations were born and died. The original genuine strips made by the unknown cosmonauts had long fallen into ruin. The Indian tribes who came after only knew of the 'gods' who had once come down from heaven by oral accounts. The priests turned the factual accounts into sacred traditions and made the people constantly build new signs for the 'gods' so that they might return one day.

As this kind of linear conjuring was unsuccessful, they began to scratch large animal figures in the ground. At first they drew all kinds of birds that were supposed to symbolise flight, but later their imagination suggested outlines of spiders, apes and fish.

Admittedly that is a hypothetical explanation of the pictures at Nazca, but surely something like it could have happened? I have seen—and anyone else can see, too—that the animal symbols and the coordinates of the landing strips are only recognisable from a great height.

But that is not all. In the vicinity of Nazca, on the rock faces, are drawings of men with rays shooting out of their heads, rather like the gloriole in paintings of Christ.

Nazca is 100 miles as the crow flies from Pisco. Suddenly I had a brainwave. Was there some connection between the trident in the Bay of Pisco, the figures on the plain of Nazca and the ruins on the plateau of Tiahuanaco? Apart from a very slight deviation, they are joined by a straight line drawn on the map. But if the plain of Nazca was a landing ground and the trident at Pisco a landing signal, then there would have to be landmarks south of Nazca too, because one can hardly imagine that all the astronauts flew in from the north, i.e. from Pisco.

And in fact large markings, whose meaning and purpose have not yet been explained, were found on high rock faces near the south Peruvian town of Mollendo, 250 miles as the crow flies from Nazca, and others continue into the deserts and mountains of the Chilean province of Antofagasta. In many places right-angles, arrows and ladders with curved rungs can be identified, or one sees whole hillsides covered with rectangles partly filled in ornamentally. All the way along the map line steep rock faces exhibit circles with rays directed inwards and ovals filled in with a chessboard pattern, while on the inaccessible hillside in the desert of Tarapacár there is a gigantic 'robot'.

On 26 August, 1968, the Chilean newspaper *el Mercurio* wrote about its discovery (a little less than 500 miles south of Nazca) under the headline 'New Archaeological Discoveries by Aerial Photography': 'A group of experts have succeeded in making a new archaeological discovery from the air. When they were flying over the desert of Tarapacár, which lies in the extreme north of Chile, they discovered the stylised figure of a man drawn in the sand. This figure is about 330 ft high, and its outline is marked with stones of volcanic origin. It is on a solitary hill about 650 ft high ... Scientific circles are of the opinion that aerial searches of this kind are of great importance for the investigation of prehistory ...'

According to the members of the expedition the 'robot' is about 330 ft high. His body is rectangular, like a chest, his legs are straight and on the thin neck sits a square head from which twelve straight antennae of equal length stick up. From hip to thigh triangular fins like the stubby wings of supersonic fighters are attached to the body on both sides.

We owe this discovery to Lautaro Nuñez of the University of the North in Chile, General Eduardo Jensen and the American Delbert Trou, who observed the ground formations closely during a flight over the desert. This genuinely sensational find was fully confirmed during a second reconnaissance flight by the Director of the Archaeological Museum of Antofagasta, Mrs Guacolda Boisset. On the heights of Pintados, she discovered and confirmed with aerial photographs a series of other stylised figures over a distance of three miles.

In the summer of 1968, the government newspaper *El*

*Arauco*, Santiago, Chile, wrote: 'Chile needs the help of someone to satisfy our chronic curiosity, for neither Gey nor Domeyko (archaeologists) has ever said anything about the platform of El Enladrillado, of which some people say that it was laid out artificially and others that it is the work of beings from another planet.'

Details about the discoveries on the plateau of El Enladrillado were made known in August, 1968. The rock-strewn plateau is about 2 miles long and at the place undestroyed by the passage of time about 870 yards wide. This terrain looks rather like an amphitheatre. If its builders were men, they must have possessed the legendary 'superhuman strength'. The stone blocks that have been moved here are rectangular, 12 to 16 ft high and 20 to 30 ft long. If giants lived here, they must still have been exceptionally big. Judging by the stone chairs, their shinbones must have been 13 ft long. No fantasy is rich enough to imagine what kind of mortals could have assembled these blocks to form an amphitheatre. The newspaper *La Mañana* of Talca, Chile, for 11.8.68 asks: 'Could this place have been a landing ground (for gods). Undoubtedly it could.' What more can one want?

The plateau of El Enladrillado can only be reached on horseback. You ride for three hours from the little village of Orto Alto de Viches to the rewarding goal at a height of 1,378 ft. The volcanic blocks that are found there in large numbers have such a smooth surface that they could only have resulted from very careful dressing. A partially interrupted landing strip, about 1,000 yards long and 65 yards wide, is also clearly recognisable on this plateau. In the neighbourhood scholars have found and still find prehistoric tools with which—presumably—the 233 geometrically shaped stone blocks, each weighing 20,000 lb, were supposed to have been dressed.

In a story dated 25.8.68 the newspaper *Concepción*, El

Sur, Chile, calls the plateau of El Enladrillado 'a mysterious place'. The site is indeed mysterious—as all sites with pre-historic traditions still are today. One's gaze turns westward over vast abysses, above which condors and eagles circle and behind which volcanoes soar up like dumb sentinels. Over towards the western hills there is a 300-foot-deep natural cave in which traces of work done by human hands are still discernible. At present scholars toy with the idea that stone-age men excavated a vein of obsidian (a glasslike forma-tion of various recent lava deposits) in order to leave behind a sample of their industrial abilities in the form of tools containing metal. I cannot quite go along with this. Stone-age men would hardly have had tools containing metal. I think that the present theory cannot be correct.

During archaeological and geological investigations scholars found a fallen monolith, which even on its side rose 6 ft from the ground. When they had turned it over with great effort, they found several faces on the other side. A puzzle worthy of inclusion in the forest of questions surrounding Easter Island.

One other remarkable find deserves to be mentioned. In the middle of the plateau stand three boulders with a diameter of 3 ft to 4 ft 6 ins. When taking measurements at the beginning of last year, it was discovered that two of these boulders are in accurate compass line from north to south. With a minimal deviation the line that runs from the first two boulders to the third cuts the horizon at the point where the sun is at its zenith in summer. Here again we must ask whether an extinct race left behind traces of astonishing astronomical knowledge or whether our ancestors were working on a 'higher task'.

Such exact evidence of the past cannot be just brushed off as coincidence.

In *El Mercurio*, Santiago, Chile, for 28.8.68, the leader of the scientific expedition, Humberto Sarnataro Bounaud, supported the point of view that an ancient unknown culture must have been at work here, because natives of the region would never have been capable of such an achievement. But, said Bounaud, somebody already knew that the plateau would make a first-class landing ground for all kinds of flying bodies. That would explain the 233 symmetrically arranged stone blocks which could have been visual signs directed skywards.

Bounaud also writes: 'Or the explanation may quite simply be that unknown beings used this place for their own ends.'

I have gone into the most recent finds on the plateau of El Enladrillado in such detail for two reasons. Firstly, because they are only known in Europe to a comparatively small circle of specialists and secondly because they fit so splendidly into my theory that the markings in the Bay of Pisco were the beginning of a straight line along which landing grounds for the cosmonauts were laid out right up to the far north of Chile.

We should always bear in mind that although the creators of ancient cultures have disappeared, the traces they left behind still question and challenge us. To find the correct answers to their questions, to meet their challenge, archaeological research institutes should receive adequate funds from their governments, but perhaps also from an international organisation, so that they can systematise and intensify their investigations. It is right and necessary for the industrial nations to invest millions of pounds in research for the future, but should research into our past be treated as the Cinderella of the present for that reason? The day may come when the nations begin an archaeological research race that they all keep a strict military secret. Then a situation will arise of the kind we experienced with the first landing on the moon, but the race that begins then will not be a question of prestige so much as a matter of cashing in.

In this connection I should like to mention a number of sites where intensive modern research could probably 'decipher' many riddles of our past in a way that would benefit technology.

The remains of a human settlement whose C-14 dating gave an age of 29,600 years were found on the island of Santa Rosa, California.

Twelve miles south of the Spanish town of Ronda, in a solitary mountain valley, lies the cave of La Pileta. It can be proved that men lived in this cave from 30,000 to 6,000 B.C. The cave walls are covered with strange stylised signs that are not just senseless scribbles, for they are executed in a masterly fashion and are often repeated. They might be a kind of writing.

In the mountains of Ennedi in the southern Sahara, Peter Fuchs found rock engravings of four female figures of a kind found nowhere else in Africa. The bodies of the figures wear clothes and exhibit tattoos similar to those found in the South Pacific. Yet the distance between the southern Sahara and the Pacific islands is 15,625 miles as the crow flies!

So-called meanders have long been known from many cave drawings in Africa and Europe. They are drawings of mazes and so far no one knows what to make of them. But now these labyrinth symbols have also been found on South American cave walls, especially in the Territorio Nacional de Santa Cruz and the Territorio de Neuguen, Argentine. Was there an exchange of ideas between the artists who created them? How else can the repetition of the same symbols be explained?

The Argentinian scholar Juan Moricz has shown that the language of the Magyars was spoken in the ancient kingdom of Quito even before the Spanish conquest. He found the same family names, the same place names and identical burial customs. When the ancient Magyars buried a dead man, they said farewell to him with the words: 'He will disappear into the constellation of the Great Bear.' And in the South American valleys of Quinche and Cochasqui there are burial mounds which are faithful reproductions of the seven main stars of Ursa Major.

A stone $7\frac{1}{2}$ ft high and 33 ft in circumference has been lying since prehistoric times on a small hill between Abancay and the River Apurimac, Peru, on the stretch Cuzco-Macchu Picchu. This Piedra de Saihuite bears reliefs which show wonderful terraces, temples and whole blocks of houses, as well as strange 'drains' and writing that is still undeciphered. Similar reliefs in this area are known by the names Rumihuasi and Intihuasi. Rumihuasi exhibits a model of a temple with a niche 4 ft 7 ins high.

In February, 1967, the well-known *National Geographic Magazine*, USA, published a story about the small tribe of the Ainus, who live on the Japanese island of Hokkaido. The Ainus still claim today with complete conviction that they are the direct descendants of 'gods' who came from the cosmos, and they repeat it in their myths.

Apollo is pictured flying over the sea on a vase which is kept in the Vatican and dates to the sixth century B.C. Apollo is sitting, playing the lyre, on a kind of shell with three long legs. The construction is drawn through the air by three massive eagle's wings.

In the park-cum-museum of Villahermosa, Tabasco (Mexico), stands a neatly carved monolith on which a snake, or rather a dragon, taking up three sides of the colossus, is depicted. Inside the animal sits a man with bent back and raised extended legs. The soles of his feet are working pedals, his left hand rests on a 'gear-lever', his right hand carries a small box. The head is enveloped in a closely-fitting helmet that covers brow, ears and chin, leaving only the face free. Directly in front of his lips is an apparatus that can be identified as a microphone. Clothing and helmet of the sitting figure fit tightly together.

On a broad copper chisel, sharpened on one side, that was found in the Royal Cemetery at Ur, one can make out, reading from top to bottom: five balls, a small box like a loudspeaker, two absolutely modern rockets, which lie next to each other and emit rays at the rear, several dragon-like figures and a pretty accurate 'copy' of the Gemini capsule. The artist who made the engravings more than 5,500 years ago must have had an enviable imagination!

Señor Gerardo Niemann (Hacienda Casa Grande, Trujillo, Peru) has two remarkable clay vessels in his private collection. One vessel is $8\frac{1}{2}$ ins high and represents a kind of 'space capsule' on which motor and exhaust are as clearly recognisable as on the relief of the rocket-driving god Kulkulkan at Palenque. A dog-like animal with gaping jaws crouches on the capsule. The second clay vessel shows a man who is using the index fingers of both hands to operate a kind of calculating machine or switchboard. This vessel is 1 ft 4 ins high. Both artefacts were found in Chicama Valley on the north coast of Peru.

No, we are not at the end, we are only at the beginning of the great discoveries pointing out of the past into the future.

# 9

# EASTER ISLAND: AN INEXHAUSTIBLE TOPIC

The remains of unknown great cultures lie on nearly all the inhabitable South Sea islands. Survivals of a completely inexplicable yet obviously very advanced technology stare the visitor mysteriously in the face and literally entice him to speculate and theorise.

Particularly on Easter Island.

We spent ten days on this tiny speck of volcanic rock in the South Pacific. The days are over when this island was only visited once every six months by a Chilean warship. We were taken there by a four-engined Lan Chile Constellation. There are no hotels there yet, so we spent the whole time in a tent. We had previously stocked up with provisions, which are scarce on the island. Twice the natives invited us to supper. We had baked salmon which they put in a hole in the earth and covered with glowing charcoal and many different kinds of leaves that form part of the secret recipes of the Rapanui housewives. We had to wait nearly two hours before the smouldering food was taken out. As a gourmet I must admit that finally palate and tongue were offered a feast that was really delicious, a pleasure that was on a par with the feast for the ears provided by the Rapanui islanders singing their folklore.

The horse is still the means of transport on the island—except for one car, which belongs to the twenty-six-year-old mayor, Ropo, who is of medium height and chubby faced, and was elected democratically by his fellow-countrymen. Ropo is the uncrowned king of the island, although there are a 'governor' and a 'police commissioner' as well. Ropo comes from an old-established family and probably knows much more about Easter Island and its unsolved puzzles than all the other islanders put together. He and two of his assistants offered to act as our guides.

The language of the Rapanui is rich in vowels: ti-ta-pe-pe-tu-ti-lo-mu. I do not speak it, so we conversed in a mixture of Spanish and English. When that failed, we tried to make ourselves understood with hands, feet and grimaces that must have been extremely funny to outsiders.

There are many accounts of the history of Easter Island and just as many theories about it. After my ten days' researches I naturally cannot say *what* took place here in the remote past, but I believe I found some arguments to show what *cannot* have taken place.

There is one theory that the ancestors of the present-day Rapanui chiselled the now world-famous statues from the hard volcanic rock during generations of arduous toil.

Thor Heyerdahl, whom I respect highly, describes in his book *Aku-Aku* how he found hundreds of stone implements lying about in confusion in the quarries. From this mass find of primitive tools Heyerdahl concluded that an unknown number of men chiselled the statues here and then precipitately abandoned their work at some time or other. They threw down their tools and left them lying where they had been working.

Using a large number of islanders who worked for eighteen days, Heyerdahl erected a medium-sized statue by means of wooden beams and a primitive but successful technique, and then moved it with the help of ropes and about a hundred men on the heave-ho principle.

Here a theory appeared to be proved in practice! Nevertheless, archaeologists all over the world protested against this example. For one thing, they said, Easter Island had always been too short of men and food to have provided the necessary number of stonemasons to carry out the enormous task—even over many generations. For another, they claimed that no finds had yet supplied proof that the islanders had ever had wood at their disposal as building material (for rollers).

After my own reflections on the spot I think I may say that the stone tool theory will not stand up for long in view of the facts, which are hard in the literal sense of the word. After Heyerdahl's successful experiment I was quite prepared to cross an unsolved puzzle off my list as solved. But when I stood in front of the lava wall in the crater Rano Raraku, I decided to let the question mark stay on my list. I measured the distance hacked away between the lava and individual statues, and found spaces of up to 6 ft over a distance of nearly 105 ft. Nobody could ever have freed such gigantic lumps of lava with small primitive stone tools.

Thor Heyerdahl made the natives hammer away for weeks with the old implements which were found in abundance. I saw the meagre result: a groove of a few inches in the hard volcanic rock! We, too, bashed away at the rock like wild men, using the biggest stones we could find. After a few hundred blows, there was nothing left of our 'tools' but a few miserable splinters, but the rock showed hardly a scratch.

The stone tool theory may be valid for some of the small statues which originated in an age nearer our own, but in my conviction and the opinion of many visitors to Easter Island it can in no case be accepted for the excavation of the raw material for the colossal statues from the volcanic stone.

The Rano Raraku crater today looks like a gigantic

sculptor's workshop in which knocking off time had been suddenly announced in the middle of work. Finished, half-finished and just begun statues lie about vertically and horizontally all over the place. Here a gigantic nose towers from the sand, there feet that no shoe could fit sprawl on the scanty grass and elsewhere a face pushes its way through as if gasping for breath.

Mayor Ropo had stood by, shaking his head, as we attacked the rock with all our might.

'What are you laughing at?' my friend Hans Neuner shouted at him. 'That's what your ancestors did, didn't they?'

Ropo gave a broad grin. With a sly look on his face, he said drily: 'So the archaeologists say.'

So far no one has been able to produce even a tolerably convincing reason why a few hundred Polynesians who found it hard enough to win their scanty nourishment took such pains to carve some 600 statues.

No one has been able to give a clue as to the highly advanced techniques with which the stone blocks were freed from the hard lava.

So far no one has been able to explain why the Polynesians (if they were the sculptors) endowed the faces with shapes and expressions for which there was no model on the island: long straight noses, narrow-lipped mouths, sunken eyes and low foreheads.

No one knows who the sculptures are supposed to represent.

Not even Thor Heyerdahl!

Perhaps, it seems presumptuous of me not only to reject Heyerdahl's theory that stone tools were used to make the statues, but to use the presence of several hundred stone implements to try to prove exactly the opposite, namely that the colossal statues could not have been made in that way.

In case that sounds incredible, here is my explanation—
as usual one that seems fantastic.

A small group of intelligent beings was stranded on
Easter Island owing to a 'technical hitch'. The stranded
group had a great store of knowledge, very advanced
weapons and a method of working stone unknown to us, of
which there are many examples around the world. The
strangers hoped they would be looked for, found and
rescued by their own people. Yet the nearest mainland was
some 2,500 miles away.

Days passed in inactivity. Life on the little island became
boring and monotonous. The unknowns began to teach the
natives the elements of speech; they told them about
foreign worlds, stars and suns. Perhaps to leave the natives
a lasting memory of their stay, but perhaps also as a sign to
the friends who were looking for them, the strangers extracted a colossal statue from the volcanic stone. Then they
made more stone giants which they set up on stone
pedestals along the coast so that they were visible from
afar.

Until suddenly and without warning salvation was there.

Then the islanders were left with a junk room of just
begun and half-finished figures. They selected the ones that
were nearest completion and year after year they hammered
doggedly away at the unfinished models with their stone
tools. But the some 200 figures that were still only sketched
on the rock face defied the 'fleabites' of the stone implements. Finally the carefree inhabitants who lived only for
the day—even today they are not very fond of hard work—
gave up the thankless task, threw away their tools and returned to their primitive caves and huts.

In other words, the arsenal of several hundred stone tools
that had failed to dint the unyielding cliff was left by them
and not by the original sculptors. I claim that the stone
tools are evidence of resignation in the face of a task that

could not be mastered.

I also suspect that the same masters gave lessons on Easter Island, at Tiahuanaco, above Sacsayhuaman, in the Bay of Pisco and elsewhere. Obviously it is only one of other possible theories and it can be opposed by referring to the great distances. But then people would be omitting to take into account my theory—and I am by no means the only one to hold it—that in the remote past there were intelligences with an advanced technology for whom the covering of vast distances in aircraft of the most varied kinds was no problem.

People may doubt my theory, but they must admit that it looks as if it had been child's play for the original sculptors to cut the stone colossi from the hard rock.

Perhaps it was only a spare time activity for them.

But perhaps they had a very specific purpose in mind.

Did they get bored with the statue game one day?

Or did they get an order that compelled them to stop.

At all events they suddenly disappeared.

So far no deep excavations have been made. Perhaps remains would be found in the lower strata that would make possible a significantly earlier dating.

The Americans are building an airfield; they are digging up the ground for a concrete runway. But I did not see any systematic excavations, nor have I heard of plans for any. The islanders—and why not indeed?—go about their business without worrying. The tourists who take the trouble to come here marvel at what they see and take souvenir snaps for the family album. Serious archaeological investigations that could clear up the puzzle are not taking place.

It is known that the Moais, as the islanders call the statues, once wore red hats on their heads and that the material for them was taken from a different quarry from the one used for heads and bodies. I have actually seen the 'hat' quarry. In comparison with the quarry in the Raro Raraku crater it is like a gravel pit dug by a child. The quarry must have been far too cramped a workplace for making the big red hats. The red hats themselves, which are brittle and porous, also make me sceptical.

Were they cut out and carved here at all?

I incline to the assumption that the red hats were cast from a mixture of gravel and red earth. Many hats are hollow inside. Did their sculptors want to save weight to make transport easier? Anyone who accepts the method of making the hats by a cast of gravel and earth—and it sounds reasonable—simultaneously has the baffling transport problem solved. The round hats must simply have been rolled from the gravel pit to the sites of the statues, which were always situated lower down.

When we discussed this possibility, Mayor Ropo thought that the hats must have been much larger when they were made in the gravel pit, because they would have lost a lot by abrasion when they were rolled down. That may be so, but even today the hats are a respectable size, with a circumference of 25 ft and a height of 7 ft 2 ins. It must still have been quite a feat to put such headgear on the heads towering 33 ft above the ground.

But why were the red hats put on the strange statues at all? So far I have not found a convincing explanation in the whole of the literature about Easter Island. So I ask myself the following questions:

Had the islanders seen 'gods' with helmets and remembered the fact when it came to making the statues?

Was that the reason why the statues did not seem complete to them without the hat-helmets?

Are they meant to express the same thing that 'helmets' and 'haloes' express on prehistoric cliffs and cave walls all over the world?

When the first white men visited Easter Island, inscribed wooden tablets still hung from the necks of the Moais, but even these first curious arrivals could not find a single islander who knew how to read the writing. So far the few wooden tablets still extant have not yielded up their secrets. Nevertheless, they are proof that the Rapanui of the past knew a script, which, I may mention in passing, is astonishingly like Chinese. The generations who came after the 'visit of the gods' forgot what the others had learnt.

Letters and inexplicable symbols are also found on the petroglyphs, the large flat stones with writing and drawings that lie scattered about on the beach like carpets. Many of these torn and fissured stones have surface areas of 24 sq yards. They lie about wherever the ground is reasonably level. On them we found fish, indefinable embryonic beings, sun symbols, balls and stars.

To make the drawings clearer to us, Mayor Ropo went over the lines with chalk. I asked him if anyone knew how to interpret the signs.

No, he said; even his father and grandfather had not been able to tell him anything about them. He himself thought that the petroglyphs contained astronomical data. He said that all the temples on the island had also been aligned according to the sun and the constellations.

Then our excursion to Easter Island paid a special dividend. Mayor Ropo took us to the beach and showed us a stone egg of astonishing proportions. While we walked round the stone relic, he explained that in Rapanui tradition this egg had originally lain in the centre of the Temple of the Sun, for the 'gods' had come to them from an egg. (Discovered at Easter, 1722, the least Easter Island could do was to produce an Easter egg as a surprise for us.) I gratefully added this information to my files on strange stone eggs all over the world.

A few yards away from the army of fallen statues, the artificial egg crumbles away on the shore of the island. Only a white catalogue number differentiates the 'egg of the gods' from the hotch-potch of stones on the beach.

# 10

# TO INDIA TO CONSULT THE SACRED
# TEXTS

'And I entered the large room that shone as brightly as the interior of a temple. Beings with human faces and human hands were running about everywhere. They were carrying all kinds of apparatus and often cases of different sizes as well. They gave them to other beings who stood behind low walls and wore peculiar headgear with the sign of the eagle. The temple hall was filled with celestial music. I did not know where it came from. Often I heard an angel's voice and once I caught the words: "Flight 101 to New York—gate 12".

'Then a cherub took me by the hand and led me to a seraph who was very kind to me and gave me a small piece of paper and said, "Your ticket". I could not decipher the divine writing on it. And then the cherub stood next to me again and led me to a big gleaming heavenly bird which stood on a large smooth area in the vast park of the heavenly beasts. The heavenly bird rested on eight black wheels that protruded from the metal belly of the motionless monster like calves' feet and seemed to be of tanned leather. The gigantic wings of the gleaming heavenly creature were spread wide. Everyone awaited the god who was to fly with us and whom my cherub called the pilot.

'As I climbed the silver ladder to the bird, I saw on the wings four great boxes, each of which had a large hole in it.

And I saw that many wheels turned in one of these holes. The heavenly bird obviously belonged to the god "Swissair", for a brightly shining wall spoke this name often.

'In the belly of the bird of the god the air was filled with the sound of harps and a pleasing smell of jasmine rose to my nostrils. Now another cherub with an incomparably lovely figure came and put me in a throne and fastened a broad band tightly round my waist. The harp music died away; a god's voice announced: "Please stop smoking and fasten your seat belts." The voice proclaimed many more prophecies which I understood as little as I had understood everything else that had been said. Suddenly a terrible noise, like the roaring and thundering of a violent storm, was heard. The bird shuddered, started to move and roared away from the other diving birds faster than the fleeing leopard. And it rushed away faster and faster, as mighty as the surge of the sea, strong as the sons of our first mother the Sun. Fear oppressed my breast like a tightly fitting red-hot band. My senses swam.

'Then the charming cherub stood beside me again, handed me intoxicating divine nectar, raised her hand and opened a sluice above me. A refreshing celestial wind blew in my face. Now I raised my eyes and lo, from out of the belly of the divine bird I could see the wings which were motionless and did not move like the flapping of birds' wings. Below me I caught sight of water and clouds and a jumble of green and brown in strange jagged shapes. I felt very disturbed and I gave a start. Then the cherub stood beside me yet again and made known to me the wisdom of the heavenly one: "Be not afraid, no one has ever stayed up here."'

I have just recounted a plane journey as it might have been told by one of our remote ancestors if he had flown from Zurich to New York in a modern jet aircraft. Apparently an absurd fancy, but we shall see that it is by no means so ridiculous.

The prophet Ezekiel (x, 1–19) gives an account that suggests a definite association of ideas after my imaginative attempt to reproduce an ancestor's story of a journey by plane:

1. 'Then I looked, and, behold, in the firmament that was above the head of the cherubims there appeared over them as it were a sapphire stone, as the appearance of the likeness of a throne.

2. And he spake unto the man clothed with linen, and said, Go in between the wheels, even under the cherub, and fill thine hands with coals of fire ... And he went in in my sight.

3. Now the cherubims stood on the right side of the house, when the man went in; and the cloud filled the inner court.

4. Then the glory of the LORD went up from the cherub, and stood over the threshold of the house; and the house was filled with the cloud, and the court was full of the brightness of the LORD's glory.

5. And the sound of the cherubims' wings was heard even to the outer court, as the voice of the Almighty God when he speaketh.

6. And it came to pass, that when he had commanded the man clothed in linen, saying, Take fire from between the wheels, from between the cherubims; then he went in, and stood beside the wheels.

9. And when I looked, behold the four wheels by the cherubims, one wheel by one cherub, and another wheel by another cherub: and the appearance of the wheels was as the colour of a beryl stone.

10. And as for their appearances, they four had one likeness, as if a wheel had been in the midst of a wheel.

11. When they went, they went upon their four sides; they turned not as they went, but to the place whither the

head looked they followed it . . .

12. And their whole body, and their backs, and their hands, and their wings, and the wheels, were full of eyes round about, even the wheels that they four had.

13. As for the wheels, it was cried unto them in my hearing, O wheel.

16. And when the cherubims went, the wheels went by them: and when the cherubims lifted up their wings to mount up from the earth, the same wheels also turned not from beside them.

17. When they stood, these stood; and when they were lifted up, these lifted up themselves also . . .

19. And the cherubims lifted up their wings, and mounted up from the earth in my sight: when they went out, the wheels also were beside them . . .'

The International Academy for Sanskrit Research in Mysore (India) was the first body to make the experiment of rendering a Sanskrit text by Maharishi Bharadvaya, a seer of an early period, in a way suited to our modern way of thinking. The result that lay before me in black and white was so astounding that during my journey to India in the autumn of 1968 I had the accuracy of the translation checked both in Mysore and at the central College of Bangalore. This is now the modern translation of an ancient Sanskrit text reads:

6. 'An apparatus that moves by inner strength like a bird, whether on earth, in the water or in the air, is called Vimana . . .

8. . . . which can move in the sky from place to place . . .

9. . . . country to country, world to world . . .

10. . . . is called a Vimana by the priests of the sciences . . .

11. . . . The secret of building flying machines . . .

12. . . . that do not break, cannot be divided, do not catch fire . . .

13. . . . and cannot be destroyed . . .

14. ... The secret of making flying machines stand still.

15. ... The secret of making flying machines invisible.

16. ... The secret of overhearing noises and conversations in enemy flying machines.

17. ... the secret of taking pictures of the interiors of enemy flying machines.

18. ... The secret of ascertaining the course of enemy flying machines.

19. ... The secret of making beings in enemy flying machines unconscious and destroying enemy machines...'

Later on in the text the thirty-one main pieces of which the machine consists are accurately described. It also enumerates sixteen kinds of metal that are needed to construct the flying vehicle, but only three of them are known to us today. All the others have remained untranslatable to date.

The experiment that was made in Mysore with this text, the age of which is still unknown, should be set up as an example of what old texts can express in modern translation.

A curiosity that leaves me no rest has always drawn me to the old Indian source books. What a mass of fascinating and mysterious information about flying machines and fantastic weapons in the remote past can be found in the translations of the Indian Vedas and epics. The Old Testament with its vigorous, vivid descriptions pales beside these Indian jewels.

My curiosity about the original sources became even greater owing to a purely chance encounter. After a lecture which I had given to a small circle in Zurich in 1963, an Indian student came up to me and said with disarming candour: 'Do you really find anything new or shocking about what you have told us? Every half-educated Indian knows the main sections of the Vedas and so knows that the gods in ancient times moved about in flying machines and possessed terrible weapons. Really, every child in India knows that!'

Basically, the nice young man only wanted to confirm my theory, and perhaps to calm me down as well, for I easily get excited about my pet subject. He achieved exactly the opposite effect.

In the years that followed I carried on a rather one-sided correspondence with Indian Sanskrit scholars. They answered my specific questions very politely and sent me photostats of Sanskrit texts that I could not read. The only people who profited by my obsession were my stamp-collecting friends. There was no peace left for me. I had to go to India—because of the texts.

In the autumn of 1968 I flew to Bangalore, the capital of the southern state of Mysore. Bangalore is the educational centre of Southern India. Yet at first I did not notice this at all. On the first day of my arrival a kaleidoscope of bewildering impressions passed before my eyes. Beggars and starvation existence—ox carts and moped taxis—women with diamonds in their noses and a red spot on their foreheads—dilapidated wooden huts and white palaces in the English colonial style—bustle in the streets and gaunt holy cows with red eyes—soldiers in bluish-green uniforms and dirty yellow water at the edges of the streets and above all the peculiar smell which seemed to penetrate right into my brain.

The University of Bangalore, which benefits from overseas aid, is magnificently equipped and full of forward-looking intellects. Professors and students work together on solving new scientific problems.

Specialist professors of Sanskrit such as Ramesh J. Patel from the Cultural Centre at Kochrab and T. S. Nandi from the University of Ahmedabad gave me their valuable time. Generally a single phone-call was enough to fix the time and place for a conversation.

I asked about the age of Vedas and epics. Scholars were unanimous in telling me that the Mahabharata, the national epic of the Indians with more than 80,000 couplets, must have originated in its first established form about 1500 B.C. But when I inquired about the original core of the epic, the answers were either 7016 or 2604 B.C. The unusual precision for datings lying so far back in the past was due to specific astronomic constellations mentioned in connection with a battle described in the Mahabharata. In spite of these astronomical data, the specialists have not yet agreed on the age of the epic. As with the Old Testament, the original author of the Mahabharata is unknown. It is suspected that a legendary figure, Vyasa, was the original creator, but it is said with considerable assurance that the last oral narrator, Sauti, also prepared the first complete written version.

For the benefit of the mathematicians who will have to feed their computers with data to find out the time dilation on interstellar flights, I may mention two numbers that I noted in Bangalore. In the Mahabharata 1,200 divine years equal 360,800 human years!

How furious I was that I could not read Sanskrit! Everyone was most helpful; I was told exactly in which texts and in which passages I should find the 'super-weapons', 'flying weapons' and 'flying machines' I was looking for. People got on the phone and warned librarians of my imminent arrival and the texts I wished to see; they gave me willing students to accompany me and make sure I found exactly what I wanted. And then when I expectantly held the answer to my questions in my hands, the essential thing was written in Sanskrit or some other Indian language. Disappointed by the meagre results, I decided to keep up the contacts I had made and return one day a wiser man.

I still had hopes that one authority would be able to satisfy my curiosity by telling me about the texts in greater detail. I had corresponded from Switzerland with Professor Dr T. S. Nandi, Sanskrit scholar in the University of Ahmedabad. I consulted him in India and through him I met Professor Esther Abraham Solomon, who is his chief. She has a vast knowledge of Sanskrit. She has been Head of the Sanskrit Department for six years and scholars throughout India look up to her as one of the greatest experts on the subject.

Ahmedabad is an old cotton town with many important mosques and tombs from the fifteenth and sixteenth centuries. It lies right on the River Sarbarmatic, has more than 1·2 million inhabitants and today is famous for the University of Gujerat founded in 1961.

Ahmedabad has a special tourist attraction, the Shaking Towers. They are the two tall minarets of a mosque, massively built and climbable inside by a spiral staircase right to the top—barefoot, of course. These towers have a peculiarity that is unique in the world. If a small group of people sets one tower in motion by a rhythmical to-and-fro movement, the other tower begins to swing too. So far the towers have easily stood up to these constant tourist antics and they look as if they will survive the leaning tower of Pisa.

Professor Nandi had arranged for me to meet Professor Esther Solomon at noon and told me: 'Go up to the first floor; her name is on the door, go in and make yourself comfortable.'

I set off in the blazing midday sun—it was November. The University was a modern functional two-storeyed limestone building with no unnecessary external trimmings. I waited in the vestibule. To a European the encouragement: 'Go in and make yourself comfortable' is very unusual. While I was waiting, I watched professors and students going into the various offices without knocking as if it was the most natural thing in the world and observed how politely and informally they mixed with each other.

Professor Solomon arrived about one o'clock. She had been kept at a seminar. She wore a simple white sari. I estimated her age at about fifty. She greeted me like an old friend, obviously because Professor Nandi had told her about me. We carried on our conversation in English and she allowed me to tape it on my portable tape-recorder.

This was our conversation: 'Professor, am I interpreting the information of your colleagues correctly if I say that Sanskrit scholars consider the old Indian Vedas and epics to be older than the Old Testament?'

'We cannot and should not make such absolute claims. Neither the ancient Indian texts nor those of the Old Testament can be dated exactly. Although we are inclining more and more to date the oldest parts of the Mahabharata to around 1500 B.C., it is a very cautious estimate and an assumption that refers to the oldest, central core of the epic. Naturally there were many additions and elaborations that were not made until "A.D.". Even today exact datings must still be made with reservations. The original nucleus of the Mahabharata may well be a hundred and more years older than 1500 B.C. You know that the oldest texts were written on the bark of palm trees, yet before these palm texts originated, the texts had already been handed down orally for many generations. There are also inscriptions on stone, but they are comparatively rare in India.'

'In your work have you come across parallels between the texts of the Old Testament and the original Indian texts?'

'Undoubtedly there are some parallels, but in my opinion these similarities can be observed in some form or other in most peoples' legends. You have only to think of an event like the Flood or the story of the gods who created men, or the heroes who were snatched up to heaven, or the constant references to the weapons they used.'

'But it is the old Indian and Tibetan texts in particular that teem with science-fiction weapons. I am thinking of

143

the divine lightning and ray weapons, of a kind of hypnotic weapon, like the one mentioned in the Mahabharata, and of the discus which the gods threw and which always returned to them like a boomerang, and of the texts that seem to be referring to bacteriological weapons. What do you think about them?'

'They are just exaggerations of fanciful descriptions of an imaginary divine power. The ancients undoubtedly felt the need to endow their leaders and kings with a mystical, mysterious nimbus. They certainly invented the incredible and invulnerable attributes later—multiplying them with each new generation.'

'Can these fantastic conceptions be reconciled with the world of ideas of primitive times?'

'Obviously. But we ourselves are always being confronted with puzzles!'

'Flying objects called Vimanas are continually being described in Indian and Tibetan texts. What do you think about them?'

'To be quite honest, I don't know what to make of them. The descriptions obviously mean something like aeroplanes, in which the gods fought in the sky.'

'Then can we or should we simply classify these traditions as myths and dispose of them like that?'

Professor Solomon thought for a moment before she answered, almost with resignation:

'Yes, we should.'

'And supposing these texts were descriptions of very remote real events?'

'That would be fantastic!'

'But would it be impossible?'

After a pause:

'I don't know, I really don't know.'

Outside I was assailed again by the intolerable heat. I strolled slowly back to the town over a bridge that seemed

Two Assyrian cylinder seals. LEFT: Two creatures, half-man, half-animal, support a third winged being. A fourth figure appears in an egg-shaped object. ABOVE: The sun, moon and spherical flying objects are again depicted on the upper edge. The object, above left, reminds us of a flying machine.

_OW: This stone ball (diameter 7 ft 1 in) stands outside a building in San José (Costa Rica) as a decoration.

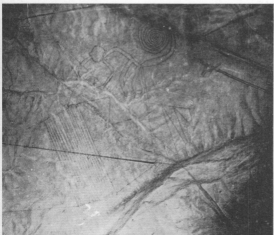

TOP: Gigantic landmark on the solitary bay south of Pisco (Peru). This vast phosphorescent sign, nearly 820 ft high, stare up at the sky.

CENTRE: If you fly over plain of Nazca, it looks li an immense air-field wit radiating and convergin landing strips. Was it on a space centre for the 'gods'?

LEFT: This ape, about 260ft high, is included i geometrical system of lines drawn with an extreme accuracy that would have been inconceivable without a knowledge of surveying

e 'hats' are hollow, with a maximum height of 7 ft 2 ins and circumferences ranging up
25 ft.

blet with writing on it from Easter Island. The script, which is still undeciphered, is not
own on any other Polynesian island.

ABOVE: The meaning and origin of this petroglyph are obscure. In the foreground a strange figure, half-fish, half-man, with a star symbol.

LEFT: The 'egg of the gods' crumbles away on the beach of Easter Island. Once a temple dedicated to the gods stood here

BELOW: This rock painting from the Central Kimberley district of Australia represents Vondjina, the mouthless mythical being of pre-history. As the personification of the Milky Way, Vondjina was the object of special reverence and given precedence over other gods.

RIGHT: These tracks run absolutely parallel and continue up the neighbouring mountainside. They link up two plateaux where there are drawings on the ground.

CENTRE: Pictures scratched on the hillsides near Nazca show figures several yards high, with radiating crowns, similar to the aureoles in Christian paintings.

BELOW: Worshipping figures in a rock drawing in Peru. According to Peruvian tradition, the zig-zag lines are an attribute of the gods.

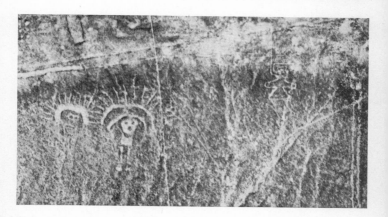

LEFT: The Dragon Monolith in the Olmec Park of Villahermosa (Mexico).

BELOW: The gigantic figures in this pictures group from Yabbaren, Tassili, are wearing space-suits. Are they men or extraterrestrial cosmonauts in antiqui

ne giants like these line the sandy shores f Easter Island. They were topped with a round stone 'hat' but these have been thrown down.

An unfinished statue on the side of the Rano Raraku crater. Is it credible that this enormous figure could have been cut out of the hard volcanic stone with primitive stone tools?

BELOW: The whole rock face is a jigsaw puzzle of uncompleted giant statues. Today two-thirds of their height is below ground.

LEFT: Typical 'Moais' physiognomy: narrow head, low brow, sunken eyes, exaggeratedly large nose, tightly compressed lips and long ears.

CENTRE: No archaeological research is going on. There is no protection of ancient monuments. So the islanders use the remains of a once powerful culture to build houses and sea-walls.

BELOW: Unfinished statues on the face of the crater Rano Raraku. The distance between the figures is 4 ft 7 ins. The men who could execute such perfect work must have possessed ultra-modern tools.

to be needless. The river had dried up to a narrow stream. Carpet makers had laid out their colourful products to dry in the river bed as far as the eye could see. Again and again I tried to recapitulate the conversation. Even this highly intelligent woman could not give a satisfactory answer to my questions.

But it is precisely what Professor Solomon could not clearly confirm that has driven me for more than a decade to compare the oldest books of mankind with my theory in mind and to track down parallels in descriptions of specific events.

Back in my hotel, the air-conditioning in my room put some life into me again. I opened the Mahabharata at random and came across this passage:

'Brighu, asked about the dimensions of the tent of the sky, answered:

"Infinite is that space inhabited by the blessed and the divinities, delightful it is, studded with many dwellings, and its boundary is unattainable.

"Above its sphere of power and below, the sun and moon are no longer seen, there the gods are their own light, shining like the sun and flashing like fire.

"And even they do not see the boundary of the mighty outspread tent of the heavens, because this is hard to reach, because it is infinite . . . But upwards and ever upwards that universe that cannot even be measured by the gods is filled with flaming, self-illuminating beings." '

The accounts in the Mahabharata still belong to the unsolved riddles of the past, even in India where this ancient text is subjected to the most minute and even pedantic scholarly scrutiny.

Ever since man has been able to think and use language, he has invented myths and legends, which, after being told for millennia, have been written down at some point in time. It is puzzling why some of these old traditions became religious or basic philosophies governing mankind's actions and others were rejected and remained without influence. A common feature of all ancient traditions is that their contents are not demonstrable, and that those which have been elevated to religions are 'believed'. When we try to interpret old texts from new standpoints today, no new versions are available to us, so we have only the old 'believed' or rejected texts to go on. Nevertheless, they provide us with startling information. But apparently it is not done to call traditionally 'believed' dogma in question or to take mythical accounts as reports of real events.

In the library of the Sorbonne in Paris I buried myself in the complete seven-volume edition of the Cabbala. Before I describe the fruits of my reading, I must say briefly that the Cabbala is certainly the most comprehensive and puzzling secret doctrine in the world. A start on writing it down is supposed to have been made around A.D. 1200. It is also said that it originated as a reaction against the realism and materialism of the Talmud.

The Cabbala interprets mysterious pronouncements in the Old Testament and comments for a circle of initiates on the encoded messages in old Jewish laws. The cabbalists say that the book was written down at God's command. It contains secret signs, symbols and mathematical formulae, and links all occult data with the mystical power of various gods. Those who belong to the small circle of initiates and have fully mastered the secrets of the Cabbala are supposed to be given the power to perform miracles.

Just as I am in the habit of considering the descriptions in other ancient texts as real, I have also taken the stories in the Cabbala as factual accounts written down after the event. It is the only way to penetrate the occult ideas of the Cabbala and find a real trace that leads from our earth to the 'gods'.

The 'seven other worlds' of the Cabbala, with their inhabitants, are described in great detail in a number of passages. Here are some extracts of which I reproduce the general sense:

'The inhabitants of the world of Geh sow and plant trees. They eat everything from trees, but know no wheat or any kind of cereal. Their world is shadowy and there are many large animals in it.

'The inhabitants of the world of Nesziah eat shrubs and plants, which they do not have to sow. They are small in stature and instead of noses have two holes in their heads through which they breathe. They are very forgetful and when doing a job of work often do not know why they began it. A red sun is seen in their world.

'The inhabitants of the world of Tziah must not eat what other beings eat. They constantly seek for underground watercourses. They are very fair of face and have more faith than all other beings. There are great riches and many handsome buildings in their world. The ground is dry and two suns are visible.

'The inhabitants of the world of Thebel eat everything from the water. They are superior to all other beings and their world is divided into zones in which the inhabitants differ facially and in colour. They make their dead to live again. The world is far away from the sun.

'The inhabitants of the world of Erez are descendants of Adam.

'The inhabitants of Adamah are also the descendants of Adam, because Adam complained of the cheerlessness of Erez. They cultivate the earth and eat plants, animals and bread. They are mostly sad and often make war on each other. There are days in this world and the groupings of the constellations are visible. In the past they were often visited by the inhabitants of the world of Thebel, but the visitors were stricken with failing memories on Adamah and no longer knew whence they came.

'The inhabitants of the world of Arqa sow and harvest. Their faces are different from our faces. They visit all worlds and speak all languages.'

Once again the old familiar questions arise. How did the authors of the Cabbala know that the beings in the seven other worlds had a different appearance from the citizens of earth? How did they know that they ate different food and knew other suns in the sky?

It is also worth mentioning the statement in the Cabbala that originally human beings did not look each other in the face during the sexual act and that the union of the seeds took place in a *single* being. Modern cabbalists claim that, before Adam, God created another being, who was 'man' only, a characteristic that did not prevent this being from producing children, who later mated with the snake.

The main work of the Cabbala, the Book Zohar, is written in Aramaic and interprets the Pentateuch from the standpoint of the cabbalistic conception of God. The Zohar is considered to be the work of Rabbi Simon bar Yochai (A.D. 130–170), but was probably first written down from oral traditions by Moses de Leon in Spain at the end of the thirteenth century and first printed in Cremona in 1558.

In the Zohar—and this is astounding—a conversation between an earthman and someone stranded from the world of Arqa is given. In this dialogue we learn that after the earth had been destroyed by fire some refugees, who had survived the catastrophe and were led by Rabbi Yossé, met a stranger, who suddenly emerged from a crack in the rock and had 'a different face'. Rabbi Yossé went up to the stranger and asked where he came from.

The stranger answered:

'I am an inhabitant of Arqa.'

The refugee was surprised and asked:

'You mean there are living creatures on Arqa?'

The stranger replied:

'Yes. When I saw you coming, I climbed out of the cave in order to learn the name of the world on which I had arrived.'

And then the stranger said that in his world the seasons were not the same as in their country; there sowing and harvest would only recommence after several years; the arrangement of the constellations was also different from the way it looked on earth...

Nearly 1,800 years of oral tradition stand behind this account, which was not written down until about 700 years ago and only printed for the first time 400 years ago. But once again I have to ask what ancient knowledge was hidden behind these words.

Naturally a stranger who visited the earth saw the constellations from a different point of view from the one he was used to on his own planet, which also had its seasons in a different sequence from the earth.

The statements are basically too realistic to be brushed aside as pure fantasy.

Then we have the Book of Dzyan with its sacred symbolic signs. No one in the world knows its real age. It is said that the original is older than the earth. It is also said that it was so strongly magnetised that the 'elect' who took it in their hands the events described pass before their eyes and at the same moment they could understand the mysterious texts through rhythmically transmitted impulses, provided their language had an adequate vocabulary.

For thousands of years this esoteric doctrine was guarded as top secret in Tibetan crypts. It was said that the secret teaching could be tremendously dangerous in the hands of ignorant people. The original text—it is not known if it still exists somewhere—was copied word for word from generation to generation and amplified by the 'elect' with new accounts and new knowledge they had acquired.

The Book of Dzyan is supposed to have originated beyond the Himalayas. By unknown routes its teaching reached Japan, India and China, and traces of its ideas are even found in South American traditions. Secret fraternities, who hid themselves in the solitary passes of the west Chinese Kun-lun range or in the deep gorges of Altyn-tag, situated also in the western part of present-day Red China, watched over collections of books of vast size. They lived in miserable temples. Subterranean vaults and passages concealed their literary treasures. The Book of Dzyan was also guarded in these fastnesses. The first fathers of the church made every effort to erase this secret doctrine from the memories of those who were familiar with it. Yet all their efforts failed and the texts were transmitted orally from generation to generation.

I have often been told about this teaching in foreign countries, but I have never met anyone who had seen a 'genuine' copy of the book. Parts of the Book of Dzyan that have been preserved or, more accurately, become known, circulate through the world in thousands of texts translated into Sanskrit. By all accounts this remarkable secret doctrine is supposed to contain the primordial ancient word, the formula of creation, and to tell of the evolution of mankind over millions of years.

The seven stanzas of the creation according to the Book of Dzyan are so interesting that I shall quote extracts from them here:

Stanza I.
'Time was not, for it lay sleeping in the infinite bosom of duration...
.. Darkness alone filled the boundless all...
... And life pulsated unconscious in universal space.
... The seven sublime lords and the seven truths had ceased to be...

Stanza II.

... Where were the builders, the luminous sons ... The producers of the form from no-form—the root of the world ... ?

... The hour had not yet struck; the ray had not yet flashed into the germ ...

Stanza III.

... The last vibration of the seventh eternity thrills through infinitude.

... The vibration sweeps along, touching with its swift wing the whole universe and the germ that dwelleth in darkness, the darkness that breathes over the slumbering waters of life ...

... The root of life was in every drop of the ocean of immortality, and the ocean was radiant light, which was fire, and heat, and motion. Darkness vanished and was no more ...

... Behold ... bright space, son of dark space ... He shines forth as the sun; he is the blazing divine dragon of wisdom.

... Where was the germ, and where now was the darkness? ...

... The germ is that and that is light, the white brilliant son of the dark hidden father.

Stanza IV.

... Listen, ye sons of the earth, to your instructors, the sons of fire ...

... Learn what we, who descend from the primordial seven, we, who are born from the primordial flame, have learnt from our fathers.

... From the effulgency of light—the ray of the ever-darkness—sprang in space the re-awakened energies ... And from the divine man emanated the forms, the sparks, the sacred animals, and the messengers of the sacred fathers...

Stanza V.

... The first seven breaths of the dragon of wisdom produce in their turn from their holy circumgyrating breaths, the fiery whirlwind.

... The swift son of the divine sons ... runs circular errands ... He passes like lightning through the fiery clouds...

... He is their guiding spirit and leader. When he commences work, he separates the sparks of the lower kingdom, that float and thrill with joy in their radiant dwellings...

Stanza VI.

... The swift and radiant one ... seats the universe on the eternal foundations...

... He builds them in the likeness of older wheels, placing them on the imperishable centres...

... How were they built by Fohat? He collects the fiery dust. He makes balls of fire, runs through them, and round them infusing life thereinto, then sets them into motion... They are cold, he makes them hot. They are dry, he makes them moist. They shine, he fans and cools them. Thus acts Fohat from one twilight to the other, during seven eternities.

... The mother's spawn filled the whole. There were battles fought between the creator and the destroyers, and battles fought for space.

Stanza VII.

... Behold the beginning of sentient formless life. First, the divine, the one from the mother spirit...

... The one ray multiplies the smaller rays...

152

... Then the builders, having donned their first clothing, descend on radiant earth and reign over men—who are themselves.'

This creation myth really needs no further commentary for the educated reader. It is uncanny how these texts really interpret themselves in the age of space travel. Only a few concepts need enlarging on.

The eternal mother = space.

Seven eternities = aeons or periods. 'Eternity' in the sense of the Christian theology is meaningless in Asiatic ideology. An aeon extends over a great period of time, i.e. a 100 years of Brahma or 311,040,000,000,000 terrestrial years. A day of Brahma consists of 4,320,000,000 years for a mortal. 'Brahma' is the force that creates and preserves all worlds. May I remind readers of the laws of time dilation without which these measurements of time are inconceivable.

Time = succession of states of consciousness.

Space = matter.

Light = something unimaginable because its source is unknown.

Father and mother = male and female principle of primordial nature.

Seven sublime lords = seven creative spirits.

Builders = the real creators of the universe, or architects of the planetary system.

Breath = dimensionless space.

Ray = matter of the world-egg.

Last vibration of the seventh eternity = periodically appearing phenomenon of the universal intelligence.

Virgin-egg = symbol of the original shape of everything visible—from the atom to heavenly bodies.

Sons of the earth—sons of fire = cosmic forces that have taken form.

Fohat = constructive force of cosmic energy.

(Quotations and exegesis are taken from *The Secret Doctrine* by Helene Petrovna Blavatsky, published in Germany in 1888!)

Other sections of the Book of Dzyan reputedly tell us that 18 million years ago there were living creatures on earth that were boneless and rubberlike, and vegetated without reason or intelligence. These beings are supposed to have reproduced themselves by division. On the course of a long evolution a peaceful and gentle kind of being originated in this way four million years ago. These beings lived in a period of tranquil bliss, in a world of happy dreamers. In the next three million years a giant race of a very different kind developed. These giants, it says in Dzyan, were androgynous and mated with themselves. Then 700,000 years ago they began to mate with 'she-animals', but terrible looking monsters resulted from this kind of reproduction. These monsters were unable to free themselves from this bestial method of reproduction and became dependent on animals and dumb like animals.

The Book of Dzyan is supposed to state that large areas of land sank in the ocean off present-day Cuba and Florida in 9,564 B.C. So far the legendary Atlantis has not been located, although the most recent theories suggest that it was the island of Thera in the Mediterranean.* Could it be identical with the sunken country that the Book of Dzyan tells about? I do not know. Perhaps Atlantis is like Unidentified Flying Objects—both of them are firmly rooted in man's imagination.

Mahabharata, Cabbala, Zohar, Dzyan. Identical as to facts that point in one direction.

Are they accounts of things that really happened?

* See *Voyage to Atlantis* by James Mavor, (Souvenir Press).

# 11

# THE PERVERSIONS OF OUR ANCESTORS

In days of yore there must have been a hybrid of man and animal. The literature and art of antiquity leaves no doubt about that. Pictures of winged beasts with human heads, mermaids, scorpion men, birdmen, centaurs and many-headed monsters are all vividly in our memory. Old books say that even in historical times these hybrid beings lived together in hordes, tribes and even in large units like peoples. They tell of specially bred hybrids, who spent their existence as 'temple animals' and seem to have been the spoilt darlings of the populace. The Sumerian kings and later the Assyrians hunted half men, half animals—possibly just for diversion. Mysterious texts refer to half-humans and hybrids, whose remarkable existence constantly evaporates into the uncheckable realms of the mythical.

The Egyptian ram still haunts the stories of the Order of Knights Templar founded in the twelfth century. He is depicted walking upright, with human hair on his head, goat's feet, goat's hind parts and a large phallus. In his *Egyptian History*, Herodotus (490–425 B.C.) speaks of strange black doves, which were supposed to have been women (II, 57). The Indian Vedas tell of mothers who 'walk on their hands'. The Epic of Gilgamesh says that Enkidu must be 'estranged from the animals'. At the wedding of Pirithoüs, the Centaurs, who had human torsoes and horses' bodies, violated the wives of the Lapithae. Six youths and six maidens had to be 'sacrificed' to the bull-headed Minotaur. Lastly it is obvious that Hephaestus's lively maidens had a sexual connotation. I also have very little doubt that the dance around the golden calf was the climax of a sexual orgy.

Plato writes in his Symposium: 'Originally there was a third sex in addition to the male and female sexes. This human had four hands and four feet ... great was the strength of these humans, their minds were presumptuous, they planned to storm heaven and to lay violent hands on the gods...'

The Cabeiri, mostly called the 'great gods' in inscriptions, carried on a mysterious cult with fertility demonds, which continued from ancient Egyptian times through the Hellenistic period and down to the flowering of the Roman culture. As the Cabeiri's rites were secret, it has not yet been possible to find out what sort of hearty sex games the ladies and gentlemen played with one another. Nevertheless, it is certain that two male and two female Cabeiri, as well as an animal, always took part in these diversions. The man and woman were not the only ones to copulate, the animal played an active role, too!

Perhaps I also ought to mention in this connection the Egyptian bulls of Apis, the 'sacred bulls of Memphis'. Because of their fertility they were mummified in sarcophagi over nine feet long and twelve feet high. Three years ago I stood in some of these musty burial chambers deep below the sands of the desert and asked myself what these fertile animals had done in their lifetime.

Tacitus (Annals, XV, 37) describes a nocturnal orgy in the house of Tigellinus at which illicit intercourse took place with the cooperation of half men, half animals. It cannot be ascertained how long these perversions were carried on in secret societies.

Sometimes the business seems to have been rather embarrassing to Herodotus and he tries to gloss it over (II, 46): 'In this province not long ago, a goat tupped a woman in full view of everybody—a most surprising incident.'

The god Pan was depicted with goat's feet and goat's head by classical artists. That, too, upset Herodotus (II, 46): 'But that is how they paint him, why, I should prefer not to mention.'

The Jewish Talmud recounts that Eve coupled with a snake. This idea inspired many artists. A woman with well-developed breasts and a snake's tail is depicted on potsherds found at Nippur—a representation, incidentally, that is not unlike those of the sirens who covet handsome young men.

Distressing as it is, the sinful side of our past cannot be wiped out. Pornography has been a sought after stimulant in all ages. Prehistoric pictures of sexual excesses on clay tablets, rock faces and animal bones speak for themselves.

Strange half human, half animal beings can be made out on the reliefs on the black obelisk of Salmanasar II in the British Museum. In the Louvre, the Museum of Baghdad and elsewhere there are pictures of remarkable couplings between animals and humans. Large stone figures with an extraordinary anatomy exist on the Island of Malta. They have globular thighs and pointed feet; they cannot be defined sexually at all. Pictures of demi-men on Assyrian works of art are no rarity. The accompanying texts tell of 'captured men-animals', who were chained, carried off and given as tribute from the land of Musri to the great king. An early stone-age bone from Le Mas d'Azil (France) shows a hybrid—half man, half ape—whose phallus must have been particularly attractive.

According to present-day biological knowledge a cross between man and animals is impossible, because the chromosome count of the partners does not tally. Such mating has never produced a viable being. But do we know the genetic code according to which the chromosome count of the mixed beings was put together?

The sexual human-animal cult which was practised with gusto and enjoyment by the people of antiquity seems to me to have been celebrated against their better judgment. Cannot the 'better judgment' of coupling with one's *own* kind have come only from unknown intelligences?

Did the earth's inhabitants lapse after the 'gods' had left?

And was this lapse the same thing as original sin?

Did they perhaps fear the day when the 'gods' would return because of their backsliding? The factor that impeded evolution in primitive times was probably interbreeding with animals. From this point of view the Fall was simply arrested or recessive evolution owing to admixture with bestial blood. 'Original sin' only becomes logical if at each birth something of the former animal side is inherited: the bestial in man.

What other kind of 'sin' was there to inherit, for heaven's sake?

The Sumerians had one single concept for the universe: an-ki, which can be roughly translated by 'heaven and earth'. Their myths tell of 'gods' who drove through the sky in barks and fireships, descended from the stars, fertilised their ancestors and then returned to the stars again. The Sumerian pantheon, the shrine of the gods, was 'animated' by a group of beings who possessed fairly recognisable human shape, but appear to have been superhuman, indeed immortal. But the Sumerian texts do not refer to their 'gods' with vague imprecision; they say quite clearly that the people had once seen them with their own eyes. Their sages were convinced that they had known the 'gods' who completed the work of instruction. We can read in Sumerian texts how everything happened. The gods gave them writing, they gave them instructions for making metal (the translation of the Sumerian word for metal is 'heavenly metal') and taught them how to cultivate barley.

We should also note that according to Sumerian records the first men are supposed to have resulted from the interbreeding of gods and the children of earth.

Sumerian tradition says that the sun god Utu and the goddess of Venus, Inanna (at least) came from the universe. The Sumerian word for rib is 'ti'; 'ti' also means 'to create life'. Ninti is also the name of the Sumerian goddess who creates life. Tradition has it that Enlil, the god of the air, made many humans pregnant. A cuneiform tablet says that Enlil discharged his seed into the womb of Meslamtaea: 'The seed of thy lord, the gleaming seed, is in thy womb; the seed of Sin, the divine name, is in my womb . . .'

Before men were created and only gods dwelt in the town of Nippur, Enlil raped the delightful Ninlil and made her pregnant on orders from above. At first the lovely child of earth, Ninlil, was unwilling to be fertilised by a 'god'. The cuneiform text from Ninlil's fear of the act of violation: 'My vagina is too small, it does not understand intercourse. My lips are too small, they do not understand how to kiss . . .'

The divine Enlil overheard Ninlil's words of refusal, but it was a decision of the 'gods' to wipe the loathsome brood of unclean beings from the face of the earth and so Enlil discharged into Ninlil's womb. On one of the tablets translated by the Sumerologist S. N. Kramer we read: 'In order to destroy the seed of mankind the decision of the council of the gods is proclaimed. According to the commanding words of An and Enlil . . . their dominion shall come to an end . . .'

So it was quite clearly a matter of wiping out the impure. Another tablet says:

'In those days, in the chamber of creation of the gods, in their house *Duku* were Lahar and Ashman formed . . .

'In those days Enki said to Enlil:

'"Father Enlil, Lahar and Ashman,

they who were created in the *Duku*,

let us allow them to descend from the *Duku*." '

Was the 'chamber of creation of the gods' the same as the 'Duku'? And was the 'Duku' from which the two were to descend the space-ship of the gods? This assumption is automatically suggested by the vivid description.

In 1889, scholars of Pennsylvania University brought home from an expedition the oldest true-to-scale town plan in the world, the plan of the town of Enlil-ki (= Nippur). In this town of the god of the air, Enlil, there was a 'gate for the sexually impure'. I think this gate was a protective measure by the gods after their work was done. After they had produced a new generation, they wanted to prevent a lapse into bestiality by separating the 'new men' from their still contaminated environment. One cuneiform tablet even gives a brief reference to the gods' method of fertilisation, namely to the implantation of the divine seed.

The Pentateuch, which has already supplied me with such a wealth of illustrative material about the means of locomotion of the galactic supermen of primitive times, is a mine of information for my theory, so long as the texts are read imaginatively, with the eyes of a man living in the age of space travel. So let us make the 'gods' come down to earth again, as it were, from Moses' descriptions. Perhaps his accounts also have something new and surprising to tell us about the practice of bestiality among primitive beings.

In Exodus, xix, 16–19, it is written:

'And it came to pass on the third day in the morning, that there were thunders and lightnings, and a thick cloud upon the mount, and the voice of the trumpets exceeding loud; so that all the people that was in the camp trembled.

'And Moses brought forth the people out of the camp to meet with God; and they stood at the nether part of the mount.

'And mount Sinai was altogether on a smoke, because the

LORD descended upon it in fire: and the smoke thereof ascended as the smoke of a furnace, and the whole mount quaked greatly.

'And ... the voice of the trumpet sounded long, and waxed louder and louder...'

Exodus xx, 18, says:

'And all the people saw the thunderings, and the lightnings, and the noise of the trumpet, and the mountain smoking: and when the people saw it, they removed, and stood afar off.'

Does anyone still believe today that the Lord God Almighty had to travel in a vehicle that smoked and flashed and caused earthquakes, and made a fiendish noise like a jet fighter? God was omnipresent. But if he was, how could he watch over and guard his 'children' and yet appear in such a terrifying way? Why did he frighten his 'children' so much that they ran away from him? The great compassionate god! Nevertheless he ordered Moses to keep the people away from the mountain where the landing took place. Exodus xix, 23–24, describes this as follows:

'... The people cannot come up to mount Sinai: for thou chargedst us, saying, Set bounds about the mount, and sanctify it.

'And the LORD said unto him, Away, get thee down, and thou shalt come up, thou, and Aaron with thee: but let not the priests and the people break through to come up unto the LORD, lest he break forth upon them.'

One of David's psalms gives a particularly dramatic description of God appearing (Psalm xxix, 7–9):

'The voice of the LORD divideth the flames of fire.

'The voice of the LORD shaketh the wilderness; the LORD shaketh the wilderness of Kadesh.

'The voice of the LORD maketh the hinds to calve and discovereth the forests...'

There is an enthusiastic account of the landing of a space-ship in Psalm civ, 3-4:

'...who maketh the clouds his chariot: who walketh upon the wings of the wind:

'Who maketh his angels spirits; his ministers a flaming fire....'

But the Prophet Micah outdoes all this (i, 3-4):

'For, behold, the LORD cometh forth out of his place, and will come down, and tread upon the high places of the earth.

'And the mountains shall be molten under him, and the valleys shall be cleft, as wax before the fire....'

Imagination needs something to spark it off. But what sparked off the chroniclers of the Old Testament? Were they describing something they had not seen at all? Time and again they implore us to believe that everything was just as they described it. And I believe them implicitly; they gave eye-witness accounts. In those days no fantasy could have given them the idea of a vehicle that spits fire, whirls up the desert sand and makes the mountains melt under it. We children of the twentieth century who have read the story of Hiroshima sense for the first time what God's epiphany as described in the Old Testament might really mean.

I should also like to examine what the Old Testament has to say about artificial insemination. 'God' (or the 'gods') had landed on earth in a cosmic vehicle. They began their most important task of fertilising the inhabitants of earth with their seed. They separated all the people chosen for this experiment from the hybrid bestial world and destined them for the 'journey into the wilderness'. There they had their guinea pigs in quarantine, so to speak. They protected them from their enemies and gave them manna and ambrosia so that they did not starve. They had to stay like that 'in the wilderness' for a whole generation. Exodus xix, 4, says:

'Ye have seen what I did unto the Egyptians, and how I bare you on eagles' wings (!), and brought you unto myself.'

If it is true that the 'gods' were masters of the genetic code, it throws light on the darkness surrounding many texts, for example the passage in Genesis i, 26–27:

'And God said, Let us make man in our image, after our likeness ... So God created man in his own image, in the image of God created he him ...'

Not until later, as I have already mentioned, was woman created from man, as Moses tells us in Genesis ii, 22:

'And the rib, which LORD GOD had taken from man, made he a woman ...'

Noah, survivor of the flood and progenitor of the races of the world, was placed in the womb of Bat-Enosh by the 'gods'. Abraham's wife Sarah, who could no longer bear children because of her advanced age, was visited by 'God' and brought her son Isaac into the world. Moses recounts this in Genesis xxi, 1:

'And the LORD visited Sarah as he had said, and the LORD did unto Sarah as he had spoken.

'For Sarah conceived, and bare Abraham a son in his old age ...'

The 'Lord' confides in the Prophet Jeremiah (i, 5):

'Before I formed thee in the belly I knew thee; and before thou camest forth out of the womb I sanctified thee ...'

The reference to knowing Jeremiah before birth is crystal clear if we take it in the sense of programming according to the genetic code. Many Old Testament stories seem to me to refer to fertilisation by the gods. In my interpretation the 'gods' created a special race to carry out the terrestrial tasks they later entrusted it with. Moses speaks about the future of the race in Genesis xv, 5:

'(God is speaking to Abraham) Look now towards heaven and tell the stars, if thou be able to number them . . . So shall thy seed be.'

But the seed must preserve its individuality, according to Leviticus xx, 24:

'. . . I am the LORD your God, which have separated you from other people.'

However, the 'gods' had no end of trouble with their creatures, who could not give up their old habit of coupling with animals. So Moses warns backsliders and threatens them with punishment in Leviticus xviii, 23 *et seq.*:

'Neither shalt thou lie with any beast to defile thyself therewith: neither shall any woman stand before a beast to lie down thereto: it is confusion.

'Defile ye not yourselves in any of these things: for in all these the nations are defiled which I cast out before you:

'And the land is defiled: therefore I do visit the iniquity thereof upon it, and the land itself vomiteth out her inhabitants.

'Ye shall therefore keep my statutes and my judgments . . .'

The punishments for sin were harsh. They had to be, for intercourse with animals was obviously an everyday occurrence. Here are examples of punishments given by Moses in Leviticus xx, 15–16:

'And if a man lie with a beast, he shall surely be put to death; and ye shall slay the beast.

'And if a woman approach unto any beast, and lie down thereto, thou shalt surely kill the woman, and the beast: they shall surely be put to death . . .'

The 'chosen people' were finally to be freed from this sexual deviation, but only after spending forty years in quarantine in the wilderness. Afterwards the new generation would be disgusted at the idea of cross-breeding with animals. So the 'gods' carried on a hard, but successful battle against the half men, half animals, and on behalf of the higher men genetically programmed by them. Consequently they only allowed the young generation to return to the 'promised land'. Listen to Numbers xiv, 29–30, on the subject:

'Your carcases shall fall in this wilderness; and all that were numbered of you, according to your whole number, from twenty years old and upward...

'Doubtless ye shall not come into the land...'

But the same strict laws also applied to life in the promised land, according to Joshua xxiii, 7 and 12–13:

'That ye come not among these nations, these that remain among you...

'If ye ... shall make marriages with them, and go into them, and they to you:

'Know for a certainty that the LORD your God will no more drive out any of these nations ... but they shall be snares and traps unto you, and scourges in your side, and thorns in your eyes...'

After the entry into the promised land, customs and mores were still very strict. Bestiality was only put an end to by new laws made by the 'gods'.

The 'gods' gave the group of men they had mutated precise hygienic instructions, which are reproduced in Leviticus xii, 2–4:

'When a man shall have in the skin of his flesh a rising, a scab, or bright spot, and it be in the skin of his flesh like the plague of leprosy; then shall he be brought unto Aaron the priest, or unto one of his sons the priests:

'And the priest shall look on the plague in the skin of the flesh: and when the hair in the plague is turned white, and the plague in sight be deeper than the skin of his flesh, it is a plague of leprosy...

'If the bright spot be white in the skin of his flesh, and in sight be not deeper than the skin, and the hair thereof be not turned white; then the priest shall shut him up that hath the plague seven days.'

'Gods', i.e. unknown intelligences, taught the new men to diagnose diseases and—as in this case—to isolate the sick.

Modern instructions for total and scrupulous disinfection were also given. The procedures are described in detail in Leviticus xv, 4–12:

'Every bed, whereon he lieth that hath the issue, is unclean: and every thing, whereon he sitteth, shall be unclean.

'And whosoever touchest his bed shall wash his clothes and bathe himself in water ...

'And he that sitteth on anything whereon he sat that hath the issue shall wash his clothes and bathe himself in water ...

'And he that touchest the flesh of him that hath the issue shall wash his clothes, and bathe himself in water ...

'And if he that hath the issue spit upon him that is clean; then shall he wash his clothes, and bathe himself in water ...

'And what saddle soever he rideth upon that hath the issue shall be unclean ...

'And whosoever toucheth any thing that was under him shall be unclean ...

'And the vessel of earth, that he toucheth which hath the issue, shall be broken ...'

Those are ultra-modern sanitary precautions. But who could have had such knowledge in antiquity? Read with my glasses—1969 model—what happened was as follows:

'Gods' came from the cosmos.

'Gods' selected a group of beings and fertilised them.

'Gods' gave the group which bore their genetic material laws and instructions for a civilisation capable of development.

'Gods' destroyed those beings who relapsed into their former ways.

'Gods' gave the chosen group an extensive knowledge of

hygiene, medicine and technology.

'Gods' imparted the art of writing and methods of cultivating barley.

I have deliberately not taken chronology into account in presenting my version. The Old Testament texts are steps in the construction of a religion; they do not reflect an accurate historical unfolding of time. Comparisons with the literature of other ancient (and older) peoples lead to the conclusion that the events chronicled in the Pentateuch could not have taken place in the period assigned to them by theologists. The Old Testament is a wonderful collection of laws and practical instructions for civilisation, of myths and bits of genuine history. This collection contains a wealth of unsolved puzzles, which religious readers have been striving to solve for centuries, but it also contains too many facts that are irreconcilable with the concept of an almighty, good and omniscient god.

The central problem is: how can an omniscient god make mistakes? Can we really call a god almighty who, after creating man, says that his work is good, but a little later is full of repentance for what he has done?

Compare Genesis i, 31:

'And God saw everything that he had made, and, behold, it was very good.'

With Genesis vi, 6:

'And it repented the LORD that he had made man on the earth, and it grieved him at his heart.'

The same god who had created man decided to destroy his work. And he did it often. Why?

The idea of 'original sin' also seems inconsistent to me. Surely god, who created man, must have known that his creations would be sinful? And if he did not know, how can he be an *omniscient* god?

God punished not only Adam and Eve for the Fall, but all their innocent descendants as well. Yet their children's children had no part in the Fall; indeed, they knew nothing about it. Did god in his rage want to be propitiated by the sacrifice of the blood of the innocent? I doubt whether an infinitely good god has feelings of revenge. Nor do I understand why almighty god later allowed his own innocent son to be put to death in a gruesome way in order to forgive the whole world for its sins.

When I ask such questions, I am not trying to denigrate or doubt the great religions. I only point out these contradictions because I am convinced that the great god of the universe has absolutely nothing in common with the 'gods' who haunt legends, myths and religions, and who affected the mutation from animal to man.

All this wealth of 'literary' evidence reminds me of a sentence with which Michel Eyquem de Montaigne (1533–1592) concluded a lecture to a circle of illustrious philosophers:

'Gentlemen, all I have done is make a bouquet from flowers already picked, adding nothing but the string to tie them together.'

Because I go into things so basically, imploring letters reach me begging me not to take the sources so literally. But our fathers were obliged to take the Bible literally for 2,000 years. If they had expressed any doubts, they would have suffered for it. Today it is permissible to discuss problems and debatable points, and so I ask more questions.

Why did 'god' and his 'angels' always show themselves in connection with phenomena such as fire, smoke, earthquakes, lightning, noise and wind? Bold and imaginative explanations are offered of the kind that can flower into axiomatic proofs in the course of 2,000 years of dialectical training. But who has the courage to take the mysterious as reality?

The Swiss Professor Dr Othmar Keel thought that these epiphanies of god ought to be understood as ideograms, in direct opposition to Professor Lindborg, who interprets the same events as hallucinatory experiences. The Old Testament scholar Dr A. Guillaume considers them to be natural events, while Dr W. Beyerlein recognises ritual parts of the Israelite religious holiday customs in nearly all the phenomena.

Scholarly explanations? I find nothing but contradictions. But the change in mental climate among the younger generation is refreshing.

Thus Dr Fritz Dumermuth wrote in the periodical of the theological faculty of Basle (No 21/1965) that 'on closer inspection the accounts in question can hardly be equated with natural phenomena of a meteorological or volcanic kind. The time has come to approach things from a new point of view if biblical research is to make any progress in explaining them.'

I suspect that the unknown intelligences did not expend their efforts on a new man purely for altruistic motives. Although it is not yet proved by research, one could assume that the 'gods' suspected the presence on earth of a material that they needed badly and that they looked for it. Was it fuel for their space-ships?

Many references point to the concluson that the 'gods' received a reward for their help in evolution. Exodus xxv, 2, mentions an offering, a concept it is easy to miss the point of. Expert German translators assured me that offerings could be taken to mean objects that were lifted up or pushed into something. This is what Moses says in Exodus xxv, 2–7:

'Speak unto the children of Israel, that they bring me an offering: of every man that giveth it willingly with his heart ye shall take my offering.

'And this is the offering which ye shall take to them; gold, and silver, and brass,

'And blue, and purple, and scarlet, and fine linen, and

goats' hair,

'And rams' skins dyed red, and badgers' skins, and shittim wood...

'Onyx stones, and stones to be set in the ephod, and in the breastplate.'

So that no mistake occurred when the offering was brought, the list was specified in detail. We find it in Numbers xxxi, 50–54:

'We have therefore brought an oblation for the LORD what every man hath gotten, of jewels of gold, chains, and bracelets, rings, earrings, and tablets...

'And all the gold of the offering that they offered up to the LORD ... was sixteen thousand seven hundred and fifty shekels.

'And Moses and Eleazar the priest took the gold ... and brought it ... for a memorial of Israel before the LORD.'

But the god of Israel would scarcely have demanded hard cash for the good he was doing his earthly children. It also appears from the text that the gifts were not intended for the priesthood, for the priests themselves had to collect and deliver the offering. In addition the proceeds of the collection for the gods was so accurately enumerated that such a niggling reckoning would be unworthy of the real god.

Was the offering the price demanded by the 'gods' for the great amount of intelligent knowledge transmitted?

The old sources give the impression that the 'gods' would not stay on our planet for ever. They carried out their plans and then disappeared for a long time. But they thought out ways of protecting the beings they had created during their absence. As they possessed extraordinary abilities, they probably made use of technology to keep a watch over them.

During the times the 'gods' were away, it was a frequent occurrence for a prophet seeking help and advice to call to his lord—as Samuel describes in Book I, iii, 1:

'And the child Samuel ministered unto the LORD before Eli. And the word of the LORD was precious in those days; there was no open vision' (i.e. god did not answer very often).

The new men were not left without protection. Texts speak of 'servants of the gods' who served on earth on higher orders, who protected the chosen ones and guarded the dwellings of the 'gods'. Were these 'servants of the gods' robots?

The Epic of Gilgamesh describes the dramatic battle between Enkidu and Gilgamesh, and the monster Chuwawa, who guarded the dwelling places of the 'gods' successfully and single-handed. The spears and clubs that Enkidu and Gilgamesh rained on him rebounded harmlessly from the 'shining monster', and behind him a 'door' spoke in the 'thunderous voice' of a human being. Clever Enkidu discovered the Achilles heel of Chuwawa, the servant of the gods, and managed to disable him.

Chuwawa was neither 'god' nor man. That emerges from a series of texts that James Pritchard published in *Ancient Near Eastern Texts* in 1950. The cuneiform text says:

'Until I have destroyed this "man", if it is man, until I have killed this god, if it is a god, I will not direct my steps to the city ... O Lord (addressing Gilgamesh), thou, who hast not seen this thing ... art not stricken with dread, I, who have seen this "man", am stricken with dread. His teeth are like dragon's teeth, his face is like a lion's face...'

Isn't that the description of a fight with a robot? Did Enkidu find out where the lever was that turned the machine off and so decide the unequal combat in his favour?

Another cuneiform translation by N. S. Kramer also makes me suspect that a 'servant of the gods' was a programmed robot:

'...those who accompanied her, the Inanna (the goddess), were beings who know no food, who know no water; eat no scattered meal, drink no offered up water...'

Sumerian and Assyrian tablets often mention such beings, who 'eat no food and drink no water'. Sometimes these weird monsters are described as 'flying lions', 'fire-spitting dragons' or 'radiant god's eggs'.

We also meet the guards left behind by the 'gods' in Greek sagas. The story of Hercules tells of the Nemean lion, who had fallen down from the moon and could not be wounded 'by any human weapon'. Another saga describes the dragon Ladon, whose eyes never slept and whose weapons were 'fire and frightful hissing'. Before Medea and Jason could carry off the Golden Fleece, they had to outwit the dragon who was covered with flashing scales of iron and who writhed about, enveloped in flames.

We also find robots in the Bible. What else could the angels have been that saved Lot and his family before Sodom and Gomorrha perished? And what are we to make of the 'arms of god' which lent a helping hand in the battles of the chosen people? Moses tells us of an angel who was helpful on god's orders in Exodus xxiii, 20–21:

'Behold, I send an angel before thee, to keep thee in the way, and to bring thee into the place which I have prepared.

'Beware of him, and obey his voice, provoke him not; for he will not pardon your transgressions: for my name is in him.'

To me it seems only logical that a robot has the name or mind of his constructor 'in him', and also that he can never deviate from his programming.

One thing I always found wonderful as a schoolboy was Jacob's experience as told in Genesis xxviii, 12. When Jacob lay down to sleep at night on one of his journeys, he saw a ladder the top of which reached to heaven and had god's angels climbing up and down it. Had Jacob perhaps surprised the 'servants of god' loading goods into a space-ship? Was Jacob's wonderful experience an eyewitness account?

As a crucial test of my audacious claims, my readers should make the experiment of reading 'robots', as we understand the term today, for dragons whenever they are mentioned in ancient texts. It is astonishing how intelligible the unintelligible suddenly becomes.

I accept the fact that the theories I have expounded will be savagely attacked. Unknown intelligences are supposed to have put an end to bestiality and unnatural sexual practices? A new species of man are supposed to have received the first instructions for a civilised communal life from intelligences? Unknown intelligences, after carrying out their task, are supposed to have vanished into the universe again, but to have left behind overseers of the new men? And these overseers are supposed to have been robots or automatons?

I try to recognise a reality that once existed behind myths, legends and traditions. Here are some incontrovertible facts.

Tibetans and Hindus called the universe the 'mother' of the terrestrial race.

The natives of Malekula (New Hebrides) state that the first race of men consisted of descendants of the 'sons of heaven'.

The Red Indians say that they are the descendants of the 'thunderbird'.

The Incas believed they descended from the 'sons of the Sun'.

The Rapanui trace their origin back to the birdmen.

The Mayas are supposed to be 'children of the Pleiades'.

The Teutons claim that their forefathers came with the 'flying *Wanen*'.

The Indians believe that they descend from Indra,

Ghurka or Bhima—all three of whom drove through the heavens in 'fireships'.

Enoch and Elijah disappeared for ever in a 'chariot of fire'.

The South Sea Islanders say they descend from the god of heaven, Tangalao, who came down from heaven in an enormous gleaming egg.

*One* core is common to these genealogical stories: 'gods' came, chose a group whom they fertilised and separated from the unclean. They imparted all kinds of modern knowledge to them and then disappeared for a period or for ever.

Karl F. Kohlenberg described what we are now left with in his book *Völkerkunde* (Ethnology):

'. . . the riddle of the gods, the riddle of the origin of man, a chaos of traditions, the real meaning of which our limited erudition is still unable to explain.'

May I be allowed one more important allusion to the 'riddle of the gods'. In my first book I mentioned the theory of relativity, the basic rocket formula and time shifts on interstellar flights. We have seen that time for the crew of a space-ship travelling just below the speed of light passes considerably slower than for those who stay behind on the launching planet. Should we regard it as coincidence that the oldest writings, quite independently of each other, constantly emphasise that the units of time applicable to the gods are different from ours?

A human generation was only a 'moment' to the Indian god Vishnu. Each of the legendary emperors of Chinese prehistory was a heavenly ruler, who drove through the sky with fire-breathing dragons and lived for 18,000 terrestrial years. Indeed P'an Ku, the first heavenly ruler, travelled around in the cosmos two million two hundred and twenty-nine thousand years ago, and even our own familiar Old Testament assures us that in the hand of God everything is 'a time and times and the dividing of time' (Daniel vii, 25), or as Psalm xc, 4, so magnificently expresses it:

'For a thousand years in thy sight are but as yesterday when it is past, and as a watch in the night.'

# 12

# QUESTIONS AND STILL MORE QUESTIONS

Were outward signs of age-old traditions misunderstood during past millennia?

Were our attempts at interpretation moving in the wrong direction?

Have we made what has always been and still is before our eyes more complicated than it really is?

Were straightforward practical and technical directions given a distorted interpretation in religious and philosophical mysteries?

Are the traditions that have accumulated in myths and religions meant to be far less mysterious and far more practical than was believed for millennia?

Will there be time for us to learn what the scanty relics of human prehistory have to tell us, before the small amount of extant material is finally devoured, scattered and destroyed by bulldozers?

When will archaeologists make a cut a kilometre long through the sandstone cliffs called the Extern Stones in the Teutoburger Forest?

When will a large-scale expedition be completely free to excavate the mystery-enshrouded sites around Marib?

When will underwater radiation investigations with modern apparatus be made in the Dead Sea?

When will archaeologists follow up the long overdue idea of sounding the Chephren Pyramid by doing the same thing under the numerous other pyramids?

When will diggers remove the top layer of Tiahuanaco so that we may learn what secrets are still hidden below it?

How long will lone wolves thirsty for knowledge have to dig in the Sahara without any help or support? When will helicopters be put at their disposal for even short periods to help them investigate the vast territory?

When will a chemical trace analysis of the Plain of Nazca finally be carried out?

For how long must amateur idealists struggle to free the ruins in the jungles of Guatemala and Honduras?

When will deep excavations be made at Zimbabwe (Southern Rhodesia)?

What international organization is ready to finance a cartographical institute that will finally clear up the specific geographical and geodesic connections that exist between the remains of mysterious primitive cultures on different continents?

Will an international organisation, possibly UNESCO, ever reach the decision to have the thousands and thousands of rock paintings and cave drawings all over the world catalogued?

Is it not possible that the keys to the 'kingdom of heaven' are hidden in many places on earth?

Were we smitten with blindness for thousands of years? And are we still?

Actually the ancient 'gods' were always telling us that we were deaf and blind, but that one day we should know the 'truth'.

From time immemorial all religions have promised that we should find the 'gods' if only we looked for them, and that once we have found them we should go to heaven and eternal peace would reign on earth.

Why should we not take this promise literally?

Perhaps we are making a mistake when we interpret the

concept 'heaven' as another-worldly, never ending state of bliss. Perhaps 'heaven' simply meant the 'universe'?

Surely we ought to seek the 'gods', and the messages they left, here on earth, instead of hoping for them somewhere in an interminable eternity?

May not these 'gods', whom mankind has longed for and prayed to in all ages, have left behind technical instructions which would enable us to meet them in the universe?

Since the beginning of human history wars have been and are being waged continuously somewhere or other on our planet. Did the 'gods' promise peace on earth because they knew that once the inhabitants of earth had felt the full impact of seeing their tiny planet from outer space they would realise that all terrestrial squabbles were utterly futile?

Do the 'gods' hope or expect that once earthly beings get to know space they will lose the national consciousness they have only assumed and instead consider the infinite cosmos as the universal motherland?

From the perspective of the universe all men will be simply inhabitants of the 'third planet', a minor sun on the edge of the galaxy—and not Russians or Chinese, Americans or Europeans, black or white.

Could mankind make its age-old dream of 'going to heaven' come true if it took up the promise of the 'gods'? That the 'gods' promised men the possibility of return to the stars is implicit in Genesis xi, 6:

'(The Lord is speaking to the people) ... and this they begin to do: and now nothing will be restrained from them, which they have imagined to do.'

And when one day the first contacts with intelligences on other planets are made, we shall soon learn to understand one another in one language as in the time of the Tower of Babel. The 2,976 languages that are spoken on earth today can still be preserved as country dialects. But scholars from all countries and on all planets will exchange their knowledge in *one* language.

But then our familiar and carefully preserved world picture will collapse and the younger generation of the space age will erase from its consciousness the last nationalistic feelings, which will have become meaningless.

For that reason alone, I think, it is our duty to examine both apparently fantastic interpretations of traditional old texts and factual stone evidence with the greatest scientific care. Once we have absorbed all the messages left behind by the 'gods', flesh and blood encounters with astronauts from distant stars will lose their terror because we shall know that these beings have something in common with us: they, too, experienced the day of their creation at some point in time.

# BIBLIOGRAPHY

ALLEN, T., *Wesen, die noch niemand sah*, Lübbe, 1966.

ANDREAS, P., and ADAMS, G., *Between Heaven and Earth*, George G. Harrap & Co., 1967.

ARDREY, R., *African Genesis*, Collins, 1961.

BLAVATSKY, H. P., *The Secret Doctrine*, Theosophical Publishing Co., 1888.

BLAVATSKY, H. P., *The Voice of Silence*, Theosophical Publishing Co., 1889.

BÖTTCHER, H. M., *Die grosse Mutter*, Econ., 1968.

BOGEN, H. J., *Modern Biology*, Weidenfeld & Nicolson, 1968.

BOSCHKE, F. L., *Creation Still Goes On*, Hodder & Stoughton, 1964.

BOSCHKE, F. L., *Die Schöpfung ist noch nicht zu Ende*, Econ, 1966.

BRAUN, W. von, *Space Frontier*, Frederick Muller, 1968.

BRENTJES, B., *Fels- und Höhlenbilder Afrikas*, Lambert Schneider, 1965.

BRION, M., *The World of Archaeology*, Elek Books, 1961-2.

CAMP, L. S. and C.C. de, *Geheimnisvolle Stätten der Geschichte*, Econ, 1966.

CHARROUX, R., *Phantastische Vergangenheit*, Herbig, 1966.

CHARROUX, R., *Verratene Geheimnisse*, Herbig, 1967.

CLARKE, A. C., *Über den Himmel hinaus*, Econ, 1960.

CLARKE, A. C., *Im höchsten Grade phantastisch*, Econ, 1967.

CLARKE, A. C., *Eine neue Zeit bricht an*, Econ, 1968.

COVARRUBIAS, M., *Indian Art of Mexico and Central America*, Knopf, New York, 1957.

DÄNIKEN, E. von, *Chariots of the Gods?*, Souvenir Press, London, 1968.

DANIELSSON, B., *Forgotten Islands of the South Seas*, Allen & Unwin, 1957.

DISSELHOLF, H.-D., *Gott muss Peruaner sein*, Brockhaus, 1956.

DISSELHOFF, H.-D., *Alltag im alten Peru*, Callwey, 1966.

EISELEY, L., *Von der Entstehung des Lebens und der Naturgeschichte des Menschen*, Fischer Bücherei, 1969.

EUGSTER, J., *Die Forschung nach ausserirdischem Leben*, Orell Füssli, 1969.

FAUST, H., *Woher wir kommen—Wohin wir gehen*, Econ, 1967.

FRISCHAUER, P., *Es steht geschrieben*, Droemer, 1967.

FUCHS, W. R., *Knaurs Buch der Denkmaschinen*, Droemer, 1968.

GAMOW, G., *Biography of the Earth*, Macmillan & Co., 1959.

GEORG, E., *Verschollene Kulturen*, Voigtländer, 1930.

GOLOWIN, S., *Götter der Atomzeit*, Francke, 1967.

GRAUL, E. H. and FRANKE, H. W., *Futurologie und Medizin*, in *Deutsches Ärzteblatt*, No. 11, 1969.

HABER, H., *Unser blauer Planet*, DVA, 1965.

HAPGOOD, CH. H., *Maps of the Ancient Sea Kings*, Chilton, Philadelphia-New York, 1965.

HEBERER, G., *Homo—unsere Ab- und Zukunft*, DVA, 1968.

HEINRICH, H., *Die ewige Weltordnung*, Hamburger Kulturverlag, 1967.

HERODOTUS, *Neun Bücher griechischer Geschichte*, Atlas, no date.

HERVÉ, A., *Charles de Carlo*, in Planet 2, 69 *et seq.*, 1969.

HEYERDAHL, T., *Aku-Aku*, Allen & Unwin, 1958.

HOENN, K., *Sumerische und akkadische Hymnen und Gebete*, Artemis, 1953.

HOMET, M. F., *Sons of the Sun*, Neville Spearman, 1963.

HONORÉ, P., *Das Buch der Altsteinzeit*, Econ, 1967.

JUNGK, R., *The Big Machine*, André Deutsch, 1969.

*Die Kabbala*, Arkana, 1962.

KAHN, H. and WIENER, A., *Ihr werdet es erleben*, Molden, 1968.

KOHLENBERG, K. F., *Völkerkunde*, Diederichs, 1968.

KOLOSIMO, P., *Not of This World*, Souvenir Press, 1970.

KRAMER, S. N., *History Begins at Sumer*, Thames & Hudson, 1958.

KRICKEBERG, W., *Märchen der Azteken und Inkaperuaner, Maya und Muisa*, Diederichs, 1928.

KRICKERBERG, W., *Altmexikanische Kulturen*, Safari, 1956.

KÜHN, H., *Wenn Steine reden*, Brockhaus, 1966.

LAJOUX, J.-D., *The Rock Paintings of Tassili*, Thames & Hudson, 1963.

LEITHÄUSER, J. G., *Inventors of Our World*, Weidenfeld & Nicolson, 1958.

LEY, W., *Die Himmelskunde*, Econ, 1967.

LHOTE, H., *Die Felsbilder der Sahara*, Zettner, 1963.

LOMMEL, A. and K., *Die Kunst des fünften Erdteils*, Catalogue of the Staatlichen Museums für Völkerkunde, Munich, 1959.

*The Mahabharata, a Criticism*, The Sanskrit Book Dept., Delhi, 1966.

*Mahabharatam*, Brockhaus, 1906.

MEISSNER, B., *Babylonien und Assyrien*, Vol. 2, Winters, 1925.

MENZEL, R., *Adam schuf die Erde neu*, Econ, 1968.

MÉTRAUX, A., *Easter Island*, André Deutsch.

NIELSEN, T., *Die letzten Geheimnisse der Erde*, Neff, 1957.

OSTEN-SACKEN, P. V. DER, *Wanderer durch Raum und Zeit*, Hirzel, 1965.

PAUWELS, L. and BERGIER, J., *Aufbruch ins dritte Jahrtausend*, Scherz, 1962.

PAUWELS, L. and BERGIER, J., *Der Planet der unmöglichen Möglichkeiten*, Scherz, 1968.

PIGGOT, S., *Die Welt aus der wir kommen*, Droemer, 1967.

PHILBERT, B., *Christliche Prophetie und Nuklearenergie*, Christiana, 1963.

PORTMANN, A., *Biologie und Geist*, Rhein Verlag, 1956.

REICHE, M., *Geheimnis der Wüste*, in Selbstverlag, Lutzweg 9, Stuttgart 80.

*Der Rig-Veda* I-IV, Harrassowitz, 1951.

RÜEGG, W., *Die ägyptische Götterwelt*, Artemis, 1959.

SÄNGER, E., *Raumfahrt heute—morgen—übermorgen*, Econ, 1964.

SÄNGER-BREDT, I., *Träumerei am Rande der Weltraumfahrt*, in Weltraumfahrt, Beiträge zur Weltraumforschung und Astronautik, Heft 2, 1954.

SAGAN, C., *Is the Early Evolution of Life Related to the Development of the Earth's Core?*, in *Nature*, Vol. 206, 1965.

SAURAT, D., *Atlantis and the Giants*, Faber & Faber, 1957.

SCHENK, G., *Die Erde—Unser Planet im Weltall*, DVA, 1965.

SHAPLEY, H., *Wir Kinder der Milchstrasse*, Econ, 1965.

SHKLOVSKY, I. S. and SAGAN, C., *Intelligent Life in the Universe*, Holden Day, San Francisco, 1966.

SIMON, P., *Noticias Historiales de las Conquistas de Tierra Firme en las Indias Occidentales*, Bogota, 1890.

STEINBUCH, K., *Falsch programmiert*, DVA, 1968.

STONE, D., *A Preliminary Investigation of the Flood Plain*, in *American Antiquity*, Vol. 9, July 1943.

STREHL, R., *Der Himmel hat keine Grenzen*, Econ, 1967.

SULLIVAN, W., *Signale aus dem All*, Econ, 1967.

TAYLOR, G. R., *The Biological Time-bomb*, Thames & Hudson, 1968.

TEILHARD DE CHARDIN, P., *Der Mensch im Kosmos*, Beck, 1965.

THIEL, R., *Der Roman der Erde*, Neff, 1959.

TRIMBORN, H., *Das alte Amerika*, Fretz & Wasmuth, 1959.

UREY, H. C., *Astrophysik III: Das Sonnensystem*, Springer, 1959.

VOGT, H. H., *Aussergalaktische Sternsysteme und die Struktur der Welt im Grossen*, Geest & Portig, 1960.

VOGT, H. H., *Das programmierte Leben*, Albert Müller, 1969.

WATSON, J. D., *The Double Helix*, Weidenfeld & Nicolson, 1968.

ZINDEL, C., *Zu den Felsbildern von Carschenna*, in Jahresbericht 1967 der Historisch-Antiquarischen Gesellschaft von Graubunden, p. 3 *et seq.*

Newspapers and periodicals (in chronological order)

'Kampfflugzeuge vor 12,000 Jahren', from *Arbeiter-Zeitung* for 17.2.63, Vienna.

'Ein Bauplan des Lebens wurde entdeckt', from *Die Zeit* for 26.3.65.

'Die Geheimschrift des Lebens ist entziffert', from *Die Zeit* for 9.7.65.

'Ein Funke Leben', from *Der Spiegel* for 18.12.67.

'Raumschiff vor 12,000 Jahren?', from *Sputnik*, Vol. 1, 1968.

'Amplia Repercusion han tenido Noticias sobre la curiosa Planicie Cordillerana', from *La Mañana* for 11.8.68, Talca, Chile.

'El Enladrillado, un Lugar de Misterio—La Frustrada Base de los OVNI', from *El Sur* for 25.8.68, Concepción, Chile.

'Neue archäologische Entdeckung', from *El Mercurio* for 26.8.68, Santiago, Chile.

'Sensacional Hallazgo en Talca—Descubren Antigua Civilización', from *Las Ultimas Noticias* for 26.10.68, Santiago, Chile.

'Das Alphabet des Lebens', from *Die Zeit* for 26.10.68.

'Ein Gehirn schwebt frei im Raum', from *Abendzeitung* for 5.4.69, Munich.

'Baum aus der Retorte', from *Süddeutsche Zeitung* for 22.4.69.

'Zusätzliches Gehirn macht klüger', from *Die Zeit* for 25.4.69.

'Der schnellste Computer in Europa', from *Süddeutsche Zeitung* for 25.4.69.

'Die massgerechten Abwehr-Moleküle', from *Die Zeit* for 25.4.69.

'Struktur eines Antikörpers aufgeklärt', from *Süddeutsche Zeitung* for 25.4.69.

'Amok im Blut', from *Der Spiegel* for 26.4.69.

'Die Sowjetsonde auf der Venus gelandet', from *Süddeutsche Zeitung* for 17/18.5.69.

'Sintflüt auf der Venus', from *Der Stern* for 1.6.69.

'Die genetische Vernichtung durch vertauschte Chromosomen', from *Die Zeit* for 23.5.69.

# ACKNOWLEDGEMENTS

Verlage B. Arthaud, Grenoble/A. Zettner, Würzburg; Verlag F. A. Brockhaus, Wiesbaden; Verlag Callwey, München; Göllner; Pressebildarchiv Hertel, Bad Berneck (Lala Aufsberg); Internationales Bildarchiv v. Irmer, München; Verlag Koehler & Amelang, Leipzig; Verlag Knopf, New York; Paolo Koch, Zürich; Herbert Kühn, Mainz; Linden-Museum, Stuttgart; H. Liszt; Andreas Lommel, München; Bildarchiv Foto Marburg; Museum für Völkerkunde, Berlin; Verlag Orbis, Prag; Maria Reiche, Stuttgart; Ullstein-Bilderdienst, Berlin; Karin Voight, Mannheim; Erich v. Däniken.

# INDEX

187

Feinberg, Gerald, 23 ff
Ferris, James, 34
Formaldehyde, 30
Fox, S. W., 34
Franke, Herbert W., 19 ff
Fuchs, Peter, 124
Fuencaliente, 84
Fuhlrott, Johann Carl, 25

Gavreau, Vladimir, 44 ff
Genetic code, 29, 35, 39, 47, 157, 163 ff
Gilgamesh, Epic of, 55, 155, 171 ff
Glycine, 30
Golfo Dulce, 88
Gravettian, 41
Gravitation, 32, 73
Graul, E. H., 98 ff
Green Bank, 17, 39
Guanine, 17, 35
Guillaume, A., 169

Hapgood, Charles, 16
Helium, 32
Herodotus, 155 ff
Hertz, 45 ff
Heyerdahl, Thor, 75 ff
Himalayas, 150
Hindus, 173
Homet, Marcel, 85
Hominids, 26, 27, 28
Honduras, 177 ff

Inanna, 159, 172
Indian balls, 87
Ingenio Valley, 116
Incas, 48, 89, 173
Intihuasi, 125
Inyo County, 84

Jensen, Gen. Eduardo, 120
Jericho, 45 ff
Jodrell Bank, 38
Johnson, Lyndon B., 37

Kahn, Herman, 36
Kantyua, 77 ff

Kassanzev, Alexander, 105 ff
Keel, Othmar, 168
Kham, 109
Kimball, A. P., 34
Kivik, 83
Kochrab, 140
Kohlenberg, Karl F., 174
Kornberg, A., 38
Kostyenki, 40
Kramer, S. N., 159, 171
Kukulkan, 126
Kun-lun, 150

La Gravette, 40 ff
La Pileta, 124
Lapithae, 156
Laussel, 40
Laven, Hannes, 95 ff
Lederberg, Joshua, 36
Le Mas d'Azil, 157
Leon, Moses de, 148
Leonardo da Vinci, 15
Lespugne, 40
Lhote, Henri, 81
Liebig, Justus von, 29
Lima, 51, 58
Lindborg, 169
Lovell, Sir Bernard, 38

Macchu-Picchu, 125
MacGowan, Roger A., 17, 20
Mahabharata, 56 ff, 141 ff
Maharishi Bharadvaya, 138
Makemake, 78
Malekula, 173
Malta, 157
Marib, 176
Martian moon theory, 108
Martinov, Dmitri, 92 ff
Matthews, 34
Mayas, 48, 89, 173
McConnell, James, 60 ff
Mendel, Gregor Johann, 33
Menhirs, 48
Methane, 29, 30
Miller, Stanley, 29 ff
Moais, 132 ff

## THE ODESSA FILE
*by* FREDERICK FORSYTH, author of THE DAY OF THE JACKAL

The life-and-death hunt for a notorious Nazi criminal unfolds against a background of international espionage and clandestine arms deals, involving rockets designed in Germany, built in Egypt, and equipped with warheads of nuclear waste and bubonic plague. Who is behind it all? Odessa. Who or what is Odessa? You'll find out in *The Odessa File* . . .

'In the hands of Frederick Forsyth the documentary thriller achieves its most sophisticated form—Mr. Forsyth has produced both a brilliant entertainment and a disquieting book.'—THE GUARDIAN

0 552 09436 6—**50p**                                            T65


## THE TERMINAL MAN
*by* MICHAEL CRICHTON

*The Terminal Man* is a novel of breathtaking suspense and alarming implications; its theme—mind control. A thrilling combination of science and fantasy, it is the story of the first operation linking a human brain and a computer.

'Michael Crichton uses literary techniques similar to those developed in his earlier novel *The Andromeda Strain*\*; but here greater impact is achieved. This is a fast, exciting novel that with its diagrams, graphs, X-ray charts and computer print-out, has all the credibility of a detailed case history or a brilliant piece of reportage.—SUNDAY TIMES
\* Also published by Corgi Books

0 552 09192 8—**50p**                                            T66

# A SELECTED LIST OF PSYCHIC, MYSTIC AND OCCULT BOOKS THAT APPEAR IN CORGI

*All these books are available at your bookshop or newsagent: or can be ordered direct from the publisher. Just tick the titles you want and fill in the form below.*

------------------------------------------------------------

CORGI BOOKS, Cash Sales Department, P.O. Box 11, Falmouth, Cornwall.
Please send cheque or postal order; no currency, and allow 10p to cover the cost of postage and packing (plus 5p each for additional copies).

NAME ...........................................................................................................

ADDRESS ....................................................................................................

(MAY 74) ...................................................................................................